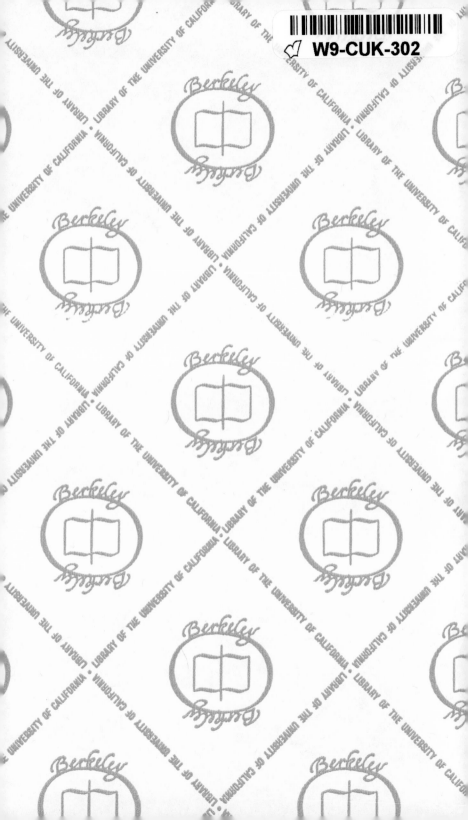

X-ray Crystallography

For Jonathan and Jessica

X-ray Crystallography

**An Introduction to the Theory and Practice of
Single-crystal Structure Analysis**

G. H. W. MILBURN, Ph.D., A.R.I.C.
Senior Lecturer,
Department of Chemistry and Biology
Sheffield Polytechnic

London   **Butterworths**

CHEMISTRY

THE BUTTERWORTH GROUP

ENGLAND
Butterworth & Co (Publishers) Ltd
London: 88 Kingsway, WC2B 6AB

AUSTRALIA
Butterworths Pty Ltd
Sydney: 586 Pacific Highway, NSW 2067
Melbourne: 343 Little Collins Street, 3000
Brisbane: 240 Queen Street, 4000

CANADA
Butterworth & Co (Canada) Ltd
Toronto: 14 Curity Avenue, 374

NEW ZEALAND
Butterworths of New Zealand Ltd
Wellington: 26–28 Waring Taylor Street, 1

SOUTH AFRICA
Butterworth & Co (South Africa) (Pty) Ltd
Durban: 152–154 Gale Street

First published 1973

© Butterworth & Co (Publishers) Ltd, 1973

ISBN 0 408 70415 2

Filmset by Photoprint Plates Ltd, Rayleigh, Essex

Printed in Great Britain by

Hazell Watson & Viney Ltd, Aylesbury, Bucks

Preface

This book has been written to provide the reader with the basic knowledge needed to solve crystal structures by X-ray diffraction methods.

It is often not realised how far ranging are the applications of X-ray methods to scientific problems generally. Any aspect of science which depends upon a knowledge of the atomic positions in a crystal can profitably make use of X-ray crystallography. The structures of proteins, metal complexes, organic molecules and geologically interesting crystals can all be subjected to an examination by X-ray methods. Parallel with the actual structure determinations is the development of the necessary computer programs which are used to facilitate the complex calculations that are carried out on the measured diffraction data. In addition, extensions and improvements to the crystallographic theory are regularly made and refinements to the methods of data collection and measurement are constantly evolving.

From this short paragraph it can be seen that X-ray diffraction methods for solving crystal structures are of interest to chemists, biologists, biochemists, physicists, geologists and mathematicians, not to mention crystallographers. The applications of crystallography are virtually limitless and open up many exciting new areas of investigation for scientists of many disciplines.

It is hoped that this book will prove useful to anyone seeking to learn of X-ray diffraction by crystals, and in particular it should be of interest to final-year science students and more specifically to first-year post-graduate crystallographers.

The subject is developed broadly along the lines of an actual structure determination, in that the earlier chapters deal with the optical examinations of crystals, the mounting of crystals, the use of X-ray cameras and the interpretation of the photographic data. Later chapters discuss the treatment of data and the use of computer programs for this purpose. Included in the text is a description of typical automatic diffractometers which can be used to facilitate greatly the collection of X-ray data, although it should not be thought that diffractometers have superseded the well tried photographic methods; an alternative to automatic data collection is automatic measurement of photographic data using a computer controlled scanner.

Acknowledgements

It would be impossible to mention by name all who have contributed to this book by discussion, suggestion, and actual work. However, I would particularly like to thank Professor M. R. Truter, who first introduced me to the subject of crystallography and very kindly helped me to maintain an interest during my military service. I must have been the only officer in the British Army following a correspondence course in X-ray crystallography!

I am indebted to the following people who gave me permission to quote from their own work: Professor D. W. J. Cruickshank, Professor M. M. Woolfson, Professor H. C. Freeman, Dr. F. R. Ahmed and Dr. P. Tollin, to whom I would like to express my grateful thanks.

I would also like to thank the manufacturers of the diffractometers, Philips Analytical Department of Pye Unicam Ltd., Hilger and Watts (the Analytical Division of Rank Precision Industries Ltd.), Siemens Ltd., and Stöe and Cie GmbH, for their help with photographs, diagrams and descriptions of their various instruments.

I thank also the Institute of Physics for permission to reproduce a picture of their Weissenberg chart, Longman's the Publishers for permission to quote from *An Introduction to Crystallography* by F. C. Phillips; and Academic Press Inc. for permission to quote from 'Image methods in crystal-structure analysis' by M. J. Buerger, which was published in *Advanced Methods of Crystallography;* and Dr. N. Bailey for the precession photograph reproduced in the book.

My thanks are also due to the members of the Sheffield Polytechnic Visual Aids Unit for their help in producing photographs suitable for publication.

I would like to express my gratitude to Professor H. Carlisle, who was the Publishers' referee, for his extremely helpful constructive criticism of the evolving manuscript. I must add, however, that any mistakes that may exist are purely my responsibility.

Finally, I would like to thank my wife for her help and encouragement during the three years this manuscript was written and for her contribution to the typing.

Contents

Part 1
Some Crystallographic Principles

1 Basic Concepts

INTRODUCTION

The inherent periodicity of crystal structure, together with the magnitude of the interatomic distances involved, e.g. a C–C single bond is 1·5445Å in length, enables a crystal to be used as a diffraction grating for an X-ray beam of wavelength comparable to interatomic distances. The solution of a crystal structure depends upon recombining, mathematically, the diffracted X-rays to synthesise an image of the molecular structure producing the diffraction. That is to say, it is possible to look at the molecules in a crystal using a mathematical lens.

A crystal structure is considered solved when it is possible to construct, from the diffraction data, a three-dimensional electron density map showing the content of the unit cell; the unit cell is the building block of the crystal lattice. This map must, of course, be chemically 'reasonable'. The molecules need not lie completely within a unit cell, and often they are found to lie across the boundaries of the cell that has been chosen, so that it is necessary to reconstruct several adjacent cells in model form to obtain the complete molecular structure. This is also necessary when looking for intermolecular contacts, e.g. hydrogen bonds.

Before commencing an investigation, much trouble can be saved if a chemical analysis of the crystals has been carried out, to ensure that the crystals under examination have the expected composition. It is assumed in this book that the starting point of a structure determination is that point where crystals of known chemical composition have been obtained and it is now desired to obtain their molecular structure.

The Crystal Lattice[1]

A crystal contains atoms arranged in a repetitive three-dimensional pattern. If each repeat unit of this pattern, which may be an atom or a group of atoms, is taken as a point then a three-dimensional point lattice is created. A space lattice such as that shown in Figure 1.1(a) is obtained when lines are drawn connecting the points of the point lattice. It can be seen that the space lattice is composed of box-like units, the dimensions of which are fixed by the lengths of the space lattice between the points in the three non-coplanar directions *a, b,* and *c.* These parallelepipeda are known as unit cells and the crystal structure has a periodicity (based on the contents of these cells) represented by the translation of the original unit of pattern along the three directions *a, b,* and *c.*

Many sets of planes can be drawn through the lattice points of a crystal structure and diffraction of X-rays by the crystal can be

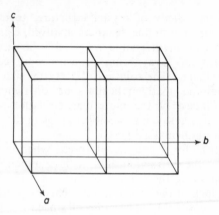

Figure 1.1(a). A space lattice

treated as reflections of the X-ray beam by these planes. It is desirable, therefore, to be able to describe each set of planes in a unique manner and this is done using Miller indices, which were originally derived to describe crystal faces but which can be applied equally well to any plane or set of planes in a crystal.

Miller Indices

Miller indices are allocated to crystal faces by first choosing a reference plane in the crystal, the parametral plane[2]. This can be any plane which has intercepts on the three crystal axes, *a, b,* and *c.* Any other plane in the crystal is then described by a/h, b/k, and c/l, where *h, k,* and *l* are small whole numbers or zero, and are known as the indices of the plane. Figure 1.1(b) shows a parametral plane with intercepts *a, b,* and *c* on the axes. Figure 1.1(c) shows a series of equidistant parallel planes parallel to the *c*-axis which is vertically upwards, and cutting the *a*- and *b*-axes. These planes are then described by their intercepts $a/1$, $b/3$, $c/0$. The Miller indices are obtained by dividing the planes intercepts into the intercepts made by the parametral plane. In this case we obtain

$$a/a/1, \quad b/b/3, \quad c/c/0$$

i.e.   1   3   0

A crystal face containing one of these planes would therefore be the (130) face which is parallel to the set of planes with interplanar spacing d_{130}.

If we apply this procedure to the parametral plane itself we obtain

$$a/a, \quad b/b, \quad c/c$$

i.e.   1   1   1

The indices of the parametral plane are always 111.

This method of allocating indices to planes is equivalent to saying that where a series of parallel equidistant planes meet an axis they

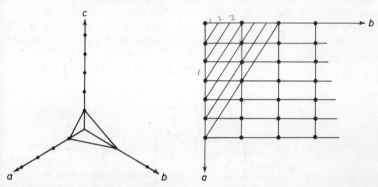

Figure 1.1 (b). A parametral plane   Figure 1.1 (c). Parallel planes

will be separated by a distance a/h, b/k, or c/l on the axes a, b, and c respectively, where hkl are the Miller indices of the plane. It follows that:

1. Crystal planes that intercept all three axes of the unit cell will have three indices hkl.
2. Planes that cut two axes and are parallel to the third axis will have indices $0kl$, $h0l$, or $hk0$ depending on whether they are parallel to axis a, b, or c, respectively.
3. Planes that cut one axis and are parallel to the other two axes will have indices $00l$, $0k0$, or $h00$ depending on whether they are parallel to axes a and b, a and c, or b and c, respectively.

It is usual to treat all X-ray reflections from planes as first order, i.e. $n = 1$ in the Bragg equation (see later), and allow for n in the indexing of the plane. Thus, a 200 reflection is treated as a first order reflection from the 200 planes, and not as a second order reflection from the 100 planes. This convention is consistent with the way in which the planes have been allocated indices and avoids much confusion.

Miller indices used to describe planes and faces are usually included in parentheses, e.g. (*hkl*); when describing reciprocal lattice points or X-ray reflections they are written without parentheses, e.g. *hkl*.

Crystal Systems

The unit cells of which a space lattice are composed do not necessarily have their three axes at right angles. The lengths of the axes can also vary from the case where they are all equal to the case where no two of them are the same. Crystals can belong to seven possible crystal systems which are characterised by the geometry of the unit cell.

Figure 1.2 shows a typical unit cell with a right-handed set of

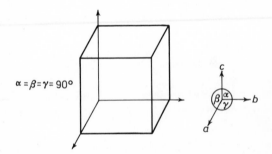

Figure 1.2. An orthogonal unit cell

axes. The cell axes and angles are labelled in the conventional way.

Table 1.1 shows the relationships of the axes and angles in each of the crystal systems.

In the discussion so far only primitive unit cells have been considered, that is the lattice points are at the corners of the unit cell. The monoclinic, orthorhombic, tetragonal and cubic systems can also have lattices where lattice points exist at the centre of faces, or the middle of the body diagonal, as well as at the corners of the unit cell. These are known as centred lattices of which there are seven and together with the seven primitive lattices they constitute the fourteen Bravais lattices[3]. Each lattice type is represented by a lattice symbol and Table 1.2 shows the possible symbols. Table 1.3 shows the distribution of the lattice types among the crystal systems.

Table 1.1

System	Axial relations	Angular relations
Triclinic	$a \neq b \neq c$	$\alpha \neq \beta \neq \gamma \neq 90°$
Monoclinic	$a \neq b \neq c$	$\alpha = \gamma = 90° \neq \beta$
Orthorhombic	$a \neq b \neq c$	$\alpha = \beta = \gamma = 90°$
Tetragonal	$a = b \neq c$	$\alpha = \beta = \gamma = 90°$
Hexagonal	$a = b \neq c$	$\alpha = \beta = 90°; \gamma = 120°$
Cubic	$a = b = c$	$\alpha = \beta = \gamma = 90°$
Trigonal i. Rhombohedral axes	$a = b = c$	$\alpha = \beta = \gamma \neq 90°, <120°$
ii. Hexagonal axes	$a = b \neq c$	$\alpha = \beta = 90°; \gamma = 120°$

Table 1.2

Lattice symbol	Lattice points
P (Primitive)	At corners
I (Inner)	At corners and centre of cell
A (Centred on two faces)	At corners and A face-centres
B (Centred on two faces)	At corners and B face-centres
C (Centred on two faces)	At corners and C face-centres
F (Centred on all faces)	At corners and all face-centres
R (Rhombohedral)	At corners

Table 1.3 LATTICE SYMBOLS

Crystal system	Possible lattices
Triclinic	P
Monoclinic	P, C
Orthorhombic	P, C, F, I
Tetragonal	P, I
Hexagonal	P
Cubic	P, F, I
Rhombohedral	R

CRYSTALS AND THEIR SYMMETRY

External Symmetry

A major landmark in the development of crystallography was the demonstration by von Laue in 1912 of the ability of crystals to diffract X-rays. Work prior to this date was necessarily concerned with the external shapes of crystals and speculation regarding their internal constitution. From 1912 onwards X-ray crystallographers have succeeded in solving increasingly complex crystal structures using a variety of methods in their approach, all of which are based on theories of crystal classification and crystal symmetry which were proposed before 1912.

Observations by the Abbé Hauy (1784) on cleavage directions in calcite led him to suggest that continual cleavage of a crystal would lead to the basic unit, from which the whole crystal may be constructed. The shape of this basic unit, according to Hauy depends upon the crystal system to which the crystal belongs. Later Hauy replaced these building units by a representative point such as a centre of gravity, joined these points to produce a lattice and thereby set the foundation on which modern theories of crystal symmetry are based.

Precise measurements of interfacial angles of crystals were made by Carangeot (1780) who developed the contact goniometer for this purpose. The reflecting goniometer, introduced by Wollaston (1809), allowed more accurate measurements to be made and led to the confirmation of the earlier detailed studies of Steno (1669), Guglielmini (1688), and de l'Isle (1772), and the confirmation of the Law of Constancy of Angle, which states that in different crystals of the same substance the values of corresponding interfacial angles will be the same.

Use can be made of the constancy of the interfacial angles in a crystal species to represent diagrammatically the symmetry and idealised shape of the crystals by means of a stereographic projection. The size and shape of a crystal face is a result of the conditions under which the crystal was grown, and as such is not a useful criterion for the description of crystal symmetry. However, because of the constancy of the interfacial angles, perpendiculars drawn from the centre of the crystal through the crystal faces are significant and the stereographic projection makes use of these perpendiculars in the following way.

A point is taken at the centre of the crystal and a sphere constructed about the crystal with this point as its centre. A series of perpendiculars may then be drawn from the centre to each face and continued to meet the surface of the sphere. The many points produced on the sphere are then projected on to the equatorial plane, by joining all the points on the northern hemisphere to the south pole and all the points on the southern hemisphere to the north pole. In the projection, the northern hemisphere points are represented by dots, and the southern hemisphere points by circles. The equatorial plane then represents the stereographic projection of the crystal.

It is useful to be able to differentiate between the accidental crystal shape that arises as a result of the conditions under which the crystal grew, and the idealised crystal shape that is demanded by the crystal's symmetry, and which may be represented on a stereographic projection. Two terms are used, crystal habit, and crystal form, and these can be defined in the following way:

Crystal habit — the incidental development of a crystal's faces that results from the conditions under which the crystal was grown.
Crystal form—a set of faces demanded by the symmetry possessed by the crystal.

A crystal may possess more than one form, and crystal habit is the modification that may occur to the various crystal forms during the crystal's growth.

The idealised external form of a crystal may be represented, as explained, by means of a stereographic projection based on a collection of normals radiating from the crystal centre to its faces. The description of the symmetry possessed by such a crystal is made in terms of a group of symmetry elements. These elements must also all act at the crystal centre, and the way in which they act will be examined in some detail.

A dictionary definition of symmetry is 'a disposition (i.e. arrangement) of parts'. In the case of crystals it is possible to describe both their internal and external symmetry in terms of a limited number of symmetry elements. The possession of symmetry by an object is usually recognised when the performance of some geometrical operation leads to self-coincidence of the object. The operation may be reflection across a plane or rotation about an n-fold axis, where $360/n$ is the number of degrees of rotation leading to self-coincidence.

The following symmetry elements are found to apply to crystals:

Rotation axes — There are five which are applicable, namely the 1-, 2-, 3-, 4-, and 6-fold rotation axes. Symbolically they are represented by the number, and n (see above) has the values 1, 2, 3, 4, 6 respectively.
Inversion axes — There are also five inversion axes, whose symbols are $\bar{1}$, $\bar{2}$, $\bar{3}$, $\bar{4}$, $\bar{6}$. An inversion axis involves rotation through $360/n$ degrees followed by inversion across a centre.
Mirror plane — A plane of symmetry or mirror plane, represented symbolically, m, reproduces on one side of itself the reflection of the other side. Alternatively it can be thought of as producing self-coincidence by reflection across a plane.

If a classification is made of known crystals by the symmetry of their external form it is found that there are thirty-two different crystal classes. Similarly, there are thirty-two possible groupings of the above symmetry elements, which are known as 'point groups'. Symmetry elements are used to describe not only crystal form and stereographic projections, but also the seven crystal systems, the fourteen Bravais lattices, and the internal symmetry of a crystal about a point such as the origin of a unit-cell. It is possible to show that by (a) combining the symmetry elements of the 14-Bravais lattices with those of the 32-point groups, or (b) by introducing two new symmetry elements, the glide-plane and the screw-axis, there results 230 space groups which describe the symmetry of all the possible internal arrangements that can be found in a crystal.

Internal Symmetry—Point Groups and Space Groups

Point groups can be used to describe the symmetry existing about a point, and use is made of them in descriptions of crystal shapes and also of atomic or molecular arrangements about a lattice point. It is the latter case which is more important to the solution of crystal structures.

In the discussion of the crystal lattice the repeat unit was considered to be a point from which was derived a three-dimensional point lattice. In a crystal the repeat motif of the three-dimensional structure will be a group of atoms or molecules which will have a symmetry relationship with the point of the point lattice. This relationship can be described in terms of a certain combination of some of the ten possible point symmetry operations. If every possible combination of point symmetry operations is taken there results the thirty-two crystallographic point groups. The symmetry operators are considered to act at the lattice point, usually the origin of the unit cell, which remains unmoved as a result of the symmetry operations. Table 1.4 shows the ten point symmetry operators and their symbols[4]. The various combinations of these ten symmetry operations are written according to certain conventions[5]:

1. The minimum number of operations to describe the symmetry are used. Others are implied.
2. The principal axis is written first followed by other axes if there are any, e.g. 42.
3. When a rotation axis has a mirror plane parallel to it, the axis is written first followed by the mirror plane, e.g. 4m.

4. A four-fold rotation axis with a mirror plane normal to it would be written $4/m$.
5. A four-fold axis with a mirror plane normal to it and mirror planes parallel to it would be written $4/mm$

i.e.

$$\frac{4}{m} . m$$

6. The same representation is applied to inversion axes.

When using symmetry elements to describe the internal arrangement of crystals, allowance must be made for translational elements of symmetry. Two such elements are used, the screw-axis and the glide plane. When taken in conjunction with the point group

Table 1.4 REPRESENTATION OF SYMMETRY ELEMENTS

Operation	Symbol
Rotation axis—two fold	2
—three fold	3
—four fold	4
—six fold	6
Inversion axis—three fold	$\bar{3}$
—four fold	$\bar{4}$
—six fold	$\bar{6}$
Plane of symmetry (mirror)	m ($\bar{2}$)
Inversion through a point	$\bar{1}$
Asymmetry	1

symmetry operations there results 230 possible space groups to describe the symmetry of crystal lattices. The operation of a glide plane combines reflection across a plane with translation parallel to the plane. The distance of translation is equal to a certain fraction of the translation that produces the line lattice.

A screw-axis combines rotation about a 2-, 3-, 4-, or 6-fold axis with translation in the direction of the axis. The translation distance is a multiple of $\frac{1}{2}$, $\frac{1}{3}$, $\frac{1}{4}$ or $\frac{1}{6}$ of the line-lattice translation.

A two-fold screw-axis would be represented 2_1 and would result in rotation about a two-fold axis and translation by $\frac{1}{2}$ the repeat distance of the axis. Table 1.5 shows the possible types of glide-planes and screw-axes[6].

In 1890 Federov and Schoenflies independently derived the 230 possible space groups that correspond to the 32 crystal classes and their symmetry operations. Towards the end of the nineteenth century papers were published by Lord Kelvin and Barlow describing three-dimensional structures in terms of lattices.

F.C. Phillips[2], describes 'the possible space groups for each crystal system by associating with every point of the Bravais lattice the elements of symmetry indicated by the crystal class, taking into account the possibilities of rotation axes in the external symmetry being represented by screw-axes in the space group, and of reflection planes in the external symmetry being represented by glide planes in the space group'.

A detailed description of all 230 space groups complete with diagrams and much other useful information is given in Volume 1 of *International Tables for X-ray Crystallography* (Kynoch Press).

As an example of the representation of a space group let us consider No. 20 in *International Tables for X-ray Crystallography*,

Table 1.5   GLIDE PLANES AND SCREW AXES

Symmetry element	Symbol	Effect
Glide plane (axial)	a	Translation $a/2$
(axial)	b	Translation $b/2$
(axial)	c	Translation $c/2$
(diagonal)	n	Translation $(a+b)/2$ or $(b+c)/2$ or $(c+a)/2$
(diamond)	d	Translation $(a\pm b)/4$ or $(b\pm c)/4$ or $(c\pm a)/4$
Screw axis	2_1	Translation $a/2$, $b/2$ or $c/2$ along axis
	3_1	Translation $c/3$
	3_2	Translation $2c/3$
	4_1	Translation $c/4$
	4_2	Translation $2c/4$
	4_3	Translation $3c/4$
	6_1	Translation $c/6$
	6_2	Translation $2c/6$
	6_3	Translation $3c/6$
	6_4	Translation $4c/6$
	6_5	Translation $5c/6$

$C222_1$. The space group symbol shows that the unit cell is C-face centred and there are two-fold rotation axes in the a and b directions with a two-fold screw axis in the c direction. In the space group $C222$, all three two-fold axes meet at the origin but in $C222_1$ the origin is at 212_1, and the two-fold axis in the b direction is at $\frac{1}{4}$ up the c-axis. Figure 1.3(a) shows the equivalent general positions of the space group. That is to say, if any one of the positions is chosen at random, the others can be produced by the application of the symmetry operations to the chosen position. The total number of general positions per unit cell is the number enclosed by the outline of the cell.

Figure 1.3(b) is a symbolic representation of the space group. For details of the symbolic representation of symmetry operators see 'symmetry element symbols' p. 27 (See also *International Tables for X-ray Crystallography*, Vol. 1.)

By convention the horizontal edge (i.e. across the page) of the unit cell is the *b*-axis direction, and the vertical direction on the page

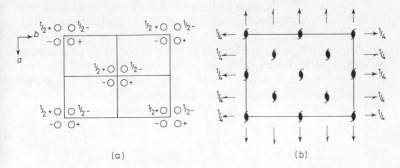

(a)   (b)

Figure 1.3(a). The general positions of space group C222₁.
(b) A symbolic representation of space group C222₁

is the *a*-axis. The *c*-axis comes upwards out of the plane of the page. The origin is at the top left-hand corner of the drawing.

GEOMETRICAL RELATIONSHIPS OF REAL AND RECIPROCAL LATTICES IN TERMS OF DIFFRACTION PATTERNS

The Reciprocal Lattice

So far we have considered only the space lattice of a crystal in real space, i.e. the real lattice. For every real lattice it is possible to construct a reciprocal lattice. (The concept of the reciprocal lattice was used by P. P. Ewald and extended by M. von Laue (1913) to describe the relationship between crystal structure and diffraction spectra.) To construct a reciprocal lattice, any point of the real lattice is taken as an origin, from which lines are drawn perpendicular to all sets of real lattice planes. Reciprocal lattice points arise on these lines at distances inversely proportional to the spacings of the real planes, and also fall on sets of parallel planes. The indices of a reciprocal lattice point, *hkl*, are the same as the indices of the planes in the real lattice that the point represents. It is important to realise that the interplanar spacing in the real lattice, and not the distance between lattice points, is the parameter which gives the distance between lattice points in reciprocal space. Figure 1.4 shows the relationship between a real two-dimensional lattice and its corresponding reciprocal lattice. A three-dimensional lattice simply consists of layers of two-dimensional lattices one above the other, the actual relationship between these layers being dependent on the crystal system.

Conventionally, the axes and angles of a unit cell in a real lattice are labelled a, b, c and α, β, γ respectively. In a reciprocal lattice they are labelled a^*, b^*, c^*, and α^*, β^*, γ^*. The volume of the cell in the real lattice is V, and in reciprocal space is V^*. The relationship between real and reciprocal unit cells depends on the particular crystal system. If $\alpha = \beta = \gamma = 90°$ then the real and reciprocal axes are coincident. In the monoclinic system where $\alpha = \gamma = 90° \neq \beta$, only the b-axis coincides with its reciprocal axis, although the

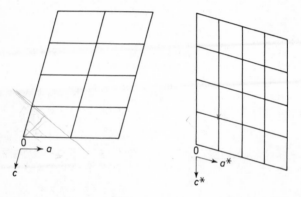

Figure 1.4. The relationship between a real and reciprocal two-dimensional lattice

a-, c-, a^*- and c^*- axes are all coplanar, as shown in Figure 1.5. In the triclinic case none of the real axes coincides with the reciprocal ones, except accidentally. Table 1.6 contains expressions for the relationships that exist between the real and reciprocal lattice parameters for the triclinic crystal system. These are simplified considerably in more symmetrical crystal systems, and a fuller list of expressions can be found in *International Tables for X-ray Crystallographers* Vol. 1, p. 13 (Kynoch Press) and in *X-ray Crystallography* by M. J. Buerger, p. 360 (John Wiley and Son).

N.B. If the reciprocal lengths obtained from photographic data are dimensionless, then a constant equal to the wavelength of the X-radiation must be introduced into the expression that is used to convert reciprocal dimensions to real dimensions.

The inverse of the above expressions are also valid, i.e. reciprocal and real values may be completely interchanged on each side of the

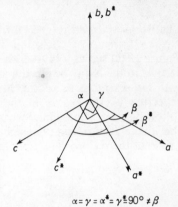

Figure 1.5. Real and reciprocal axes of the monoclinic system with the b-axis unique. The axes c, a, c*, a* are all in the same plane

$$\alpha = \gamma = \alpha^* = \gamma^* \neq 90° \neq \beta$$

expressions, e.g. in Table 1.6 equation 1 becomes $a^* = (bc \sin \alpha/V)$, etc.

The basic relationship between real and reciprocal unit cell lengths is implicit in the definition of the reciprocal lattice.

i.e.
$$a^* = K/d_{100}$$
$$b^* = K/d_{010}$$
$$c^* = K/d_{001}$$

d, the interplanar spacing, may or may not equal the real axis length depending upon the crystal system. K is a constant. If a^*, b^* and

Table 1.6 CELL RELATIONSHIPS

1. $a = \dfrac{b^* c^* \sin \alpha^*}{V^*}$; $b = \dfrac{a^* c^* \sin \beta^*}{V^*}$; $c = \dfrac{a^* b^* \sin \gamma^*}{V^*}$

2. $V = abc\sqrt{(1 - \cos^2 \alpha - \cos^2 \beta - \cos^2 \gamma + 2 \cos \alpha \cos \beta \cos \gamma)}$

3. $V = 1/V^*$

4. $\cos \alpha = \dfrac{\cos \beta^* \cos \gamma^* - \cos \alpha^*}{\sin \beta^* \sin \gamma^*}$

5. $\cos \beta = \dfrac{\cos \alpha^* \cos \gamma^* - \cos \beta^*}{\sin \alpha^* \sin \gamma^*}$

6. $\cos \gamma = \dfrac{\cos \alpha^* \cos \beta^* - \cos \gamma^*}{\sin \alpha^* \sin \beta^*}$

c^* are obtained in a dimensionless form, from X-ray photographic data, then $K = \lambda$, the radiation wavelength used in the data collection. Otherwise, $K = 1$; d is the interplanar spacing in real space. The subscript refers to the particular set of planes.

The Reciprocal Lattice and Diffraction of X-rays

Figure 1.6(a) shows a crystal, the rotation axis of which is perpendicular to the X-ray beam. If a sphere of unit radius is constructed with the crystal at its centre, X-ray diffraction by the crystal (i.e. reflection by the planes in the crystal) can be described in the following way. The point B where the primary beam passes out of the sphere after passing through the crystal is considered to be the origin of the reciprocal lattice. As the crystal rotates about its axis,

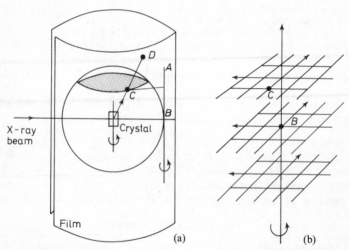

Figure 1.6(a). The reflecting sphere and a diffracted ray. AB is the rotation axis of the reciprocal lattice. Only one reciprocal lattice point is shown, C, which is on the first layer. (b) A rotating lattice. Three layers of a reciprocal lattice. The points B and C in Figure 1.6(a) are shown

the reciprocal lattice rotates about an axis through B, which is parallel to the crystal rotation axis. Whenever a point on the reciprocal lattice, e.g. C, cuts the surface of the sphere a reflected X-ray beam will travel from the crystal, through the lattice point on the sphere and register on, for instance, a photographic film at D which may be in the form of a cylinder coaxial with the crystal rotation axis, i.e. as with a Weissenberg camera.

Figure 1.6(b) shows a side view of the rotating lattice. Each layer contains a two-dimensional net of lattice points that will sweep out a circular section across the sphere (one such section is shown hatched on Figure 1.6(a). If all the points on the circumference of this section are joined to the crystal then a cone of diffracted rays is created for each layer of the reciprocal lattice. These cones correspond to the cones of diffraction discussed later, but in the present case they have been derived by means of the reciprocal lattice. For a proof that a diffracted beam is emitted whenever a reciprocal lattice point cuts the sphere, 'the sphere of reflection', see *Chemical Crystallography* p. 157, by C. W. Bunn (Oxford).

The description of the diffraction of X-rays by a crystal in terms of the reciprocal lattice and the sphere of reflection has important applications in the interpretation of photographic data. For example, it is possible to allocate a set of *hkl* indices to every spot on a Weissenberg photograph on the assumption that the photograph represents a distorted picture of the reciprocal lattice. In other words, each spot on the film can be related to the set of planes which produced it. Similarly, a precession photograph presents an undistorted representation of part of the reciprocal lattice.

FOURIER TRANSFORM

In a chapter dealing with basic concepts something should be said about Fourier transform theory, although at this point it is a little premature to involve ourselves with the detailed theory of Fourier syntheses which are dealt with in a later chapter.

The whole purpose of undertaking an X-ray crystal structure analysis is to obtain an image of the material doing the X-ray scattering, by some suitable recombination of the scattered rays. The formation of the image can be considered to proceed in two steps: the scattering of the radiation by the object, and the recombination of the scattered rays to form the image. If an optical analogy is considered, where the object doing the scattering is viewed under a microscope, it can be seen that there is no problem in recombining the scattered rays as they will automatically have the correct phase relationships. In the case of X-rays the recombination can be accomplished by means of a Fourier transformation.

The three dimensional-periodic electron density in a crystal can be represented by the three dimensional Fourier series:

$$\rho(x, y, z) = \sum_{h'} \sum_{k'} \sum_{l'} C_{h'k'l'} \, e^{2\pi i(h'x + k'y + l'z)} \tag{1.1}$$

where h^1, k^1, l^1 are integers with values from $-\infty$ to $+\infty$, and x, y, z are fractions of the period related to a set of axes.

The structure factor, F_{hkl}, which is also dealt with in more detail later on, can be considered as the resultant of j waves scattered in the direction of the reflection hkl by the j atoms in the unit cell, or alternatively as the sum of the wavelets scattered from all the elements of electron density in the unit cell, when no assumption needs to be made regarding the distribution of the density. The structure factor is then given in the latter case by the expression

$$F_{hkl} = \int_v \rho(x, y, z)\, e^{2\pi i(hx + ky + lz)}\, dV \qquad (1.2)$$

compared with

$$F_{hkl} = \sum_j f_j\, e^{2\pi i(hx_j + ky_j + lz_j)} \qquad (1.3)$$

of the former case. Combination of equations (1.1) and (1.2) leads to an expression which contains a periodic exponential

$$F_{hkl} = \int_v \sum_{h^1} \sum_{k^1} \sum_{l^1} C_{h^1 k^1 l^1}\, e^{2\pi i[(h + h^1)x + (k + k^1)y + (l + l^1)z]}\, dV$$

The integral over one period is zero except where $h^1 = -h, k^1 = -k$, and $l^1 = -l$, when the periodicity disappears and:

$$F_{hkl} = V C_{\overline{hkl}}$$

rearranging

$$C_{\overline{hkl}} = \frac{1}{V}\, F_{hkl}$$

Equation (1.1) then becomes:

$$\rho(x, y, z) = \frac{1}{V} \sum_h \sum_k \sum_l F_{hkl}\, e^{-2\pi i(hx + ky + lz)} \qquad (1.4)$$

Equations (1.3) and (1.4) describe the relationship of Fourier transformation. The former expression gives the structure factors in terms of the electron density in real space, the latter expression gives the electron density in real space in terms of the structure factors in reciprocal space. The structure factors are described as the Fourier transform of the electron density, and the converse is also true.

The solution of crystal structures by Fourier methods requires a knowledge of the relative phases of the scattered rays. An alternative approach of great importance in the study of disordered structures such as fibres, is to find an arrangement of atoms that can be shown to produce the diffraction pattern that has been observed. An optical solution to the problem may be obtained, since the diffraction of X-rays by a group of atoms can be simulated by the diffraction of light from a group of holes in an opaque mask. Considerable advances have been made in this branch of crystallography. The relevant procedures are described by C. A. Taylor and H. Lipson in *Optical Transforms* (Bell, 1964) based on the development of earlier work of W. L. Bragg. (A. Klug and D. J. de Rosier* have described the use of the Taylor–Lipson Diffractometer for image retrieval from electron micrographs of tobacco mosaic virus, and the helical tail of a bacteriophage.)

The total scattering from a group of N atoms, each having a scattering factor $f_n(s)$ and co-ordinate r_n is given by

$$G(s) = \sum_{n=1}^{n=N} f_n(s) \exp 2\pi i r_n . S$$

r_n relates an atom n to an arbitrarily chosen origin. The vector S relates the directions of the incident and scattered rays. Both r_n and S exist in three dimensions.

The above expression can be considered to give the total scattering relative to an electron at the origin, or equally the scattering from a mask of holes of transparency distribution f_n and position r_n relative to a unit scatterer of infinitesimal extent at the origin. The expression above is the Fourier transform of the atomic distribution or of the arrangement of holes.

It is difficult to represent three-dimensional objects optically, but by restricting the values of r_n to those in the plane normal to the incident beam, projections of crystal structures may be treated as masks of holes, and comparisons may be made between the corresponding S planes (having values of S only in a plane normal to the incident beam) in the optical and X-ray diffraction patterns.

The Fourier transform of the set of atoms is closely related to the diffraction pattern produced by the set of holes, although the diffraction pattern does not define the Fourier transform completely. To do this both the phase and the amplitude would be needed at each point and the diffraction pattern yields only one, the intensity. To differentiate between the Fourier transform and the optical diffraction pattern, the latter is termed the optical transform. It is in fact possible to carry out optical Fourier syntheses for non-centrosymmetric projections and to prepare full three-dimensional representations†.

The fascinating fields of protein and virus structure are ones in which use may be made of the correlation of electron micrographs

Nature, **212,** 29 (1966)
†HARBURN, G., and TAYLOR, C. A., *Proc. Roy. Soc.,* A, **264,** 339 (1961)

with X-ray diffraction data. The general principles of image reconstruction can be seen for example in papers by A. Klug and D. J. de Rosier, *Nature*, **217**, 130 (1968); and J. T. Finch and A. Klug, *J. Mol. Biol.*, **24**, 289 (1967).

Unfortunately, to look at the subject in any greater detail would take us beyond the scope of this book.

CHOICE OF RADIATION

The radiation emitted by an X-ray tube is characteristic of the target material. Three components of the emitted radiation are of use crystallographically, these are K_{α_1}, K_{α_2}, and K_β. The most common target material is probably copper, although molybdenum is often used with diffractometers. Figure 1.7 shows the spectrum of radiation

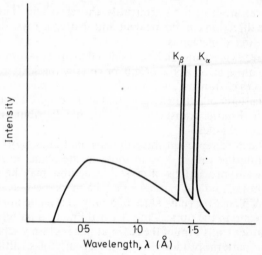

Figure 1.7. X-ray spectrum obtained from a copper target. A nickel filter removes the CuK$_\beta$ radiation

obtained from a copper target. The CuK$_\beta$ radiation can be filtered out by allowing the X-ray beam to pass through a thin sheet of nickel foil[7]. A foil 0·021 mm thick reduces the β/α radiation ratio by a factor of 600, and usually a piece 0·001 mm thick serves as a suitable filter[8].

The position of a diffracted spot on a film is a function of the wavelength used and at high θ values the wavelengths of CuK$_{\alpha_1}$ (1·5405 Å) and CuK$_{\alpha_2}$ (1·5443 Å) are sufficiently different to allow two spots to occur on the film for each set of planes. At low θ values the two spots are coincident. The intensity of these spots is in the ratio $\alpha_1/\alpha_2 : 2/1$ and when intensities are being measured the values for each spot are added together. Use can be made of the $\alpha_1 - \alpha_2$ separation at high θ values to calculate unit cell dimensions[9].

Certain elements, if present in the crystal, will absorb specific wavelengths from the incident radiation, and a suitable target must be chosen so that this does not occur. The higher the atomic number of the target the shorter the wavelength of the emitted radiation. K_β radiation is highly absorbed by an element 1 or 2 atomic numbers less than it, and K_α radiation is highly absorbed by an element 2 or 3 atomic numbers less than it. Elements that are immediately higher to the target material in the atomic table are transparent to the radiation. M. J. Buerger in *X-ray Crystallography*, p. 177, gives a table from which suitable target material can be selected for a large variety of diffracting elements.

POWDER PHOTOGRAPHS AND *d* SPACINGS[10]

Figure 1.8 shows the arrangement of a film about a rotating crystalline powder, and the sort of diffraction pattern that arises.

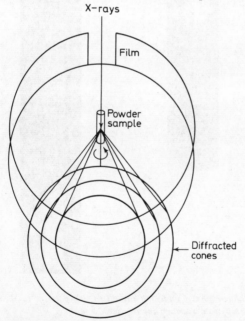

Figure 1.8. The formation of arcs on a powder photograph by the cones of diffracted X-rays

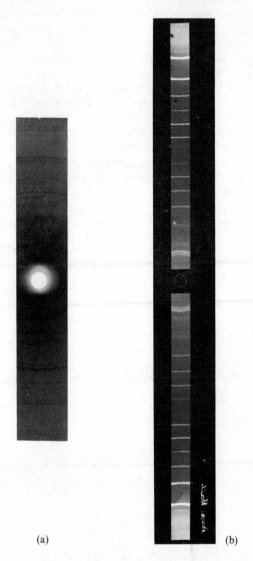

(a)   (b)

Figure 1.9. Photographs of typical X-ray diffraction patterns (a) From a grainy tin powder sample; (b) from a sodium chloride powder sample

Where the cone cuts the film it will trace out an arc. The geometry of the apparatus is such that $\theta = S/4R$ where R is the camera radius, S is the distance between corresponding lines each side of the direct beam, and θ is in radians. $n\lambda = 2d \sin \theta$ and d is given by one of the expressions in Table 1.7 depending on the crystal system. The following relationship can be derived in the case of the cubic system:

$$\sin^2 \theta = \frac{n^2\lambda^2}{4a_0^2}(h^2 + k^2 + l^2)$$

from which a_0 can be determined, the length of the cell edge.

The diffracted cone of reflections arises from the many possible

Table 1.7 EQUATIONS FOR d SPACINGS

Monoclinic:	$d = 1 \left/ \left(\dfrac{\dfrac{h^2}{a^2} + \dfrac{l^2}{c^2} - \dfrac{2hl \cos \beta}{ac}}{\sin^2 \beta} + \dfrac{k^2}{b^2} \right)^{\frac{1}{2}} \right.$
Orthorhombic:	$d = 1 \left/ \left(\dfrac{h^2}{a^2} + \dfrac{k^2}{b^2} + \dfrac{l^2}{c^2} \right)^{\frac{1}{2}} \right.$
Hexagonal and rhombohedral:	$d = 1 \left/ \left(4/3 . \dfrac{(h^2 + hk + k^2)}{a^2} + \dfrac{l^2}{c^2} \right)^{\frac{1}{2}} \right.$
Tetragonal:	$d = 1 \left/ \left(\dfrac{(h^2 + k^2)}{a^2} + \dfrac{l^2}{c^2} \right)^{\frac{1}{2}} \right.$
Cubic:	$d = a/(h^2 + k^2 + l^2)^{\frac{1}{2}}$

reflecting planes presented by the powder to the incident beam. The crystals are usually finely ground and the specimen is rotated during the exposure to X-rays to prevent graininess in the diffraction pattern. Figures 1.9(a) and (b) show typical powder photographs.

ILLUSTRATIVE TABLES AND DIAGRAMS

UNIT CELL SYMMETRY

Crystal system	Lattice symmetry
Triclinic	$\bar{1}$
Monoclinic	$2/m$
Orthorhombic	mmm
Tetragonal	$4/mmm$
Rhombohedral	$\bar{3}m$
Hexagonal	$6/mmm$
Cubic	$m3m$

BRAVAIS LATTICE SYMMETRY

	Lattice type	Symmetry
Cubic	*P*	*Pm*3*m*
	I	*Im*3*m*
	F	*Fm*3*m*
Tetragonal	*P*	*P*4/*mm*
	I	*I*4/*mm*
Orthorhombic	*P*	*Pmmm*
	C	*Cmmm*
	I	*Immm*
	F	*Fmmm*
Monoclinic	*P*	*P*2/*m*
	C	*C*2/*m*
Triclinic	*P*	$P\bar{1}$
Trigonal	*R*	$R\bar{3}m$
Hexagonal	*C* (or *P*)	*P*6/*mmm*

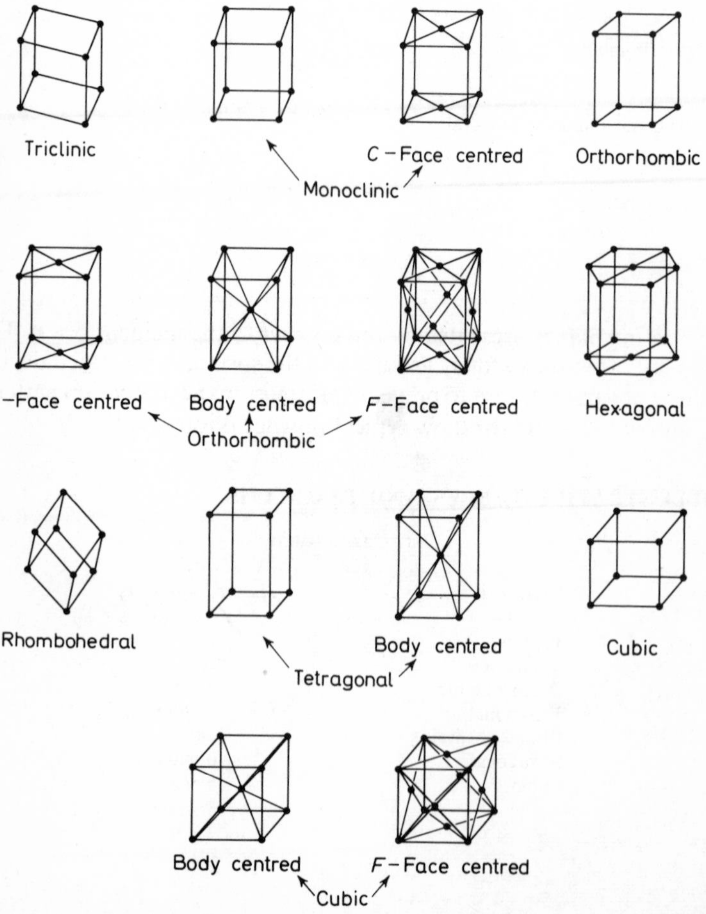

Figure 1.10. The 14 Bravais lattices

System	Point groups	Space groups						
Triclinic	1	$P1$						
	$\bar{1}$	$P\bar{1}$						
Monoclinic	2	$P2$	$P2_1$	$C2$				
	m	Pm	Pc	Cm	Cc			
	$2/m$	$P2/m$	$P2_1/m$	$C2/m$	$P2/c$	$P2_1/c$	$C2/c$	
Orthorhombic	222	$P222$	$P222_1$	$P2_12_12$	$P2_12_12_1$	$C222_1$	$C222$	$F222$
		$I222$	$I2_12_12_1$					
	$mm2$	$Pmm2$	$Pmc2_1$	$Pcc2$	$Pma2$	$Pca2_1$	$Pnc2$	$Pmn2_1$
		$Pba2$	$Pna2_1$	$Pnn2$	$Cmm2$	$Cmc2_1$	$Ccc2$	$Amm2$
		$Abm2$	$Ama2$	$Aba2$	$Fmm2$	$Fdd2$	$Imm2$	$Iba2$
		$Ima2$						
	mmm	$Pmmm$	$Pnnn$	$Pccm$	$Pban$	$Pmma$	$Pnna$	$Pmna$
		$Pcca$	$Pbam$	$Pccn$	$Pbcm$	$Pnnm$	$Pmmn$	$Pbcn$
		$Pbca$	$Pnma$	$Cmcm$	$Cmca$	$Cmmm$	$Cccm$	$Cmma$
		$Ccca$	$Fmmm$	$Fddd$	$Immm$	$Ibam$	$Ibca$	$Imma$
Tetragonal	4	$P4$	$P4_1$	$P4_2$	$P4_3$	$I4$	$I4_1$	
	$\bar{4}$	$P\bar{4}$	$I\bar{4}$					
	$4/m$	$P4/m$	$P4_2/m$	$P4/n$	$P4_2/n$	$I4/m$	$I4_1/a$	
	422	$P422$	$P42_12$	$P4_122$	$P4_12_12$	$P4_222$	$P4_22_12$	$P4_322$
		$P4_32_12$	$I422$	$I4_122$				
	$4mm$	$P4mm$	$P4bm$	$P4_2cm$	$P4_2nm$	$P4cc$	$P4nc$	$P4_2mc$
		$P4_2bc$	$I4mm$	$I4cm$	$I4_1md$	$I4_1cd$		
	$\bar{4}2m$	$P\bar{4}2m$	$P\bar{4}2c$	$P\bar{4}2_1m$	$P\bar{4}2_1c$	$P\bar{4}m2$	$P\bar{4}c2$	$P\bar{4}b2$
		$P\bar{4}n2$	$I\bar{4}m2$	$I\bar{4}c2$	$I\bar{4}2m$	$I\bar{4}2d$		
	$4/mmm$	$P4/mmm$	$P4/mcc$	$P4/nbm$	$P4/nnc$	$P4/mbm$	$P4/mnc$	$P4/nmm$
		$P4/ncc$	$P4_2/mmc$	$P4_2/mcm$	$P4_2/nbc$	$P4_2/nnm$	$P4_2/mbc$	$P4_2/mnm$
		$P4_2/nmc$	$P4_2/ncm$	$I4/mmm$	$I4/mcm$	$I4_1/amd$	$I4_1/acd$	
Trigonal	3	$P3$	$P3_1$	$P3_2$	$R3$			
	$\bar{3}$	$P\bar{3}$	$R\bar{3}$					
	32	$P312$	$P321$	$P3_112$	$P3_121$	$P3_212$	$P3_221$	$R32$
	$3m$	$P3m1$	$P31m$	$P3c1$	$P31c$	$R3m$	$R3c$	
	$\bar{3}m$	$P\bar{3}1m$	$P\bar{3}1c$	$P\bar{3}m1$	$P\bar{3}c1$	$R\bar{3}m$	$R\bar{3}c$	
Hexagonal	6	$P6$	$P6_1$	$P6_5$	$P6_2$	$P6_4$	$P6_3$	
	$\bar{6}$	$P\bar{6}$						
	$6/m$	$P6/m$	$P6_3/m$					
	622	$P622$	$P6_122$	$P6_522$	$P6_222$	$P6_422$	$P6_322$	
	$6mm$	$P6mm$	$P6cc$	$P6_3cm$	$P6_3mc$			
	$\bar{6}m2$	$P\bar{6}m2$	$P\bar{6}c2$	$P\bar{6}2m$	$P\bar{6}2c$			
	$6/mmm$	$P6/mmm$	$P6/mcc$	$P6_3/mcm$	$P6_3/mmc$			
Cubic	23	$P23$	$F23$	$I23$	$P2_13$	$I2_13$		
	$m3$	$Pm3$	$Pn3$	$Fm3$	$Fd3$	$Im3$	$Pa3$	$Ia3$
	432	$P432$	$P4_232$	$F432$	$F4_132$	$I432$	$P4_332$	$P4_132$
		$I4_132$						
	$\bar{4}3m$	$P\bar{4}3m$	$F\bar{4}3m$	$I\bar{4}3m$	$P\bar{4}3n$	$F\bar{4}3c$	$I\bar{4}3d$	
	$m3m$	$Pm3m$	$Pn3n$	$Pm3n$	$Pn3m$	$Fm3m$	$Fm3c$	$Fd3m$
		$Fd3c$	$Im3m$	$Ia3d$				

STEREOGRAPHIC PROJECTION SYMMETRY

Triclinic	Monoclinic and orthorhombic	Trigonal	Tetragonal	Hexagonal	Cubic
1	2	3	4	6	23
$\bar{1}$	m ($\bar{2}$)	$\bar{3}$	$\bar{4}$	$\bar{6}$	$\bar{2}3 = 2/m3$
$1/m = \bar{2}$	2/m	$3/m = \bar{6}$	4/m	6/m	m3 (2/m3)
$1m = \bar{2}$	mm (2m)	3m	4mm	6mm	2m3 = 2/m3
$\bar{1}m = 2/m$	$\bar{2}m = 2m$	$\bar{3}m$	$\bar{4}2m$	$\bar{6}m2$	$\bar{4}3m$
12 = 2	222	32	42	62	43 (432)
1/mm = 2m	mmm (2/mm)	$3/mm = \bar{6}m$	4/mmm	6/mmm	m3m (4/m3m)

The 32 crystal classes

SYMMETRY ELEMENT SYMBOLS

Symmetry element	Symbol	Representation: Parallel to projection plane	Perp. to projection plane
Centre	$\bar{1}$		
2-fold axis	2		
3-fold axis	3		
4-fold axis	4		
6-fold axis	6		
2-fold screw axis	2_1		
3-fold screw axis	3_1		
3-fold screw axis	3_2		
4-fold screw axis	4_1		
4-fold screw axis	4_2		
4-fold screw axis	4_3		
6-fold screw axis	6_1		
6-fold screw axis	6_2		
6-fold screw axis	6_3		
6-fold screw axis	6_4		
6-fold screw axis	6_5		
mirror	m		
a glide plane	a		
b glide plane	b		
c glide plane	c		
n glide plane	n		
d glide plane	d	$3/8$ $1/8$	

Laue Symmetry

The symmetry of diffraction effects from a crystal at first glance may be thought to reflect the point group symmetry of the crystal under study. However, the diffraction pattern differs from the point group symmetry in one important respect; it always has a centre of symmetry even when one is not present in the crystal, since reflections hkl and $\bar{h}\bar{k}\bar{l}$ are identical. This is Friedel's law and is discussed later.

The diffraction pattern symmetry is therefore that of the crystal point group together with a centre of symmetry. In this way the 32 point groups, when applied to diffraction patterns, reduce to 11 centro-symmetric point groups or Laue groups.

Crystal system	Laue symmetry
Triclinic	$\bar{1}$
Monoclinic	$2/m$
Orthorhombic	mmm
Tetragonal	$4/m$; $4/mmm$
Trigonal and hexagonal	$\bar{3}$; $\bar{3}m$
	$6/m$; $6/mmm$
Cubic	$m3$; $m3m$

(See also Table 2.4)

The significance of this can be seen by considering the monoclinic system. In this system there are three possible point groups 2; m; and $2/m$. The symmetry of the X-ray diffraction pattern produced by each of these point groups would be $2/m$; the three groups could not be distinguished from one another by their diffraction pattern.

2 Some Practical Considerations

CHOOSING A CRYSTAL

The data collected from a crystal are only as good as the quality of the crystal. It pays, therefore, to spend some time in ensuring that a good crystal has been selected. Before an optical examination is made of the crystals it is necessary to calculate the optimum size of the crystal when allowance is made for the effects of absorption on the diffracted rays[11]. Absorption corrections to intensity data are mentioned later, but some account of the reasons for allowing for absorption by the crystal when a suitable crystal is being chosen, should be mentioned.

It is assumed that when X-ray diffraction data is collected from a crystal, the crystal is completely bathed in the X-ray beam so that all parts of the crystal are exposed to the same radiation intensity. This places a limit on the size of the crystal, as its cross-section should be less that the cross-section of the X-ray beam which is approximately 0·7 mm in diameter.

A further restriction in crystal size is then introduced by the absorption effects of the crystal on the diffracted X-rays. On the other hand it is desirable to choose a crystal which is as large as possible, as the intensity of the diffracted ray is proportional to the amount of scattering material, i.e. the crystal volume.

The problem of absorption effects arises when systematic errors are introduced due to the incident and diffracted rays for different reflections passing through different volumes of the crystal and being attenuated by different amounts. There is therefore an optimum size for a crystal which is a function of the linear absorption coefficient. Ideally, the crystal should be a sphere of diameter equal to or less than the optimum size. In practice the shape of the crystal may vary widely from the ideal and, therefore, in addition to using a crystal which is within the required size limit it is also necessary to apply absorption corrections to the measured intensity data as described in a later chapter. The optimum size is calculated as follows:

If μ is the linear absorption coefficient, $2/\mu$ gives the optimum size. The value of μ is obtained from the following expression:

$$\mu = d \sum_i P_i(\mu_i/\rho)$$

d = density of the crystal in g/cm^3

P_i = fraction of each element in the compound

μ_i/ρ = mass absorption coefficients of each element in the compound for the particular wavelength of radiation being used.

If cis-Pt(NH$_3$)$_2$Cl$_2$ [12] is taken as an example, the following results are obtained, using the values for the mass absorption coefficients given in *International Tables for X-ray Crystallography* Vol. III p. 162–165.

Element	Atomic weights	Fractional composition	μ_i/ρ	$P \cdot \mu_i/\rho$
Pt × 1	195·09	0·65	200·0	130·00
N × 2	28·014	0·09	7·52	0·70
Cl × 2	70·906	0·24	106·0	25·01
H × 6	6·048	0·02	0·435	0·01
				155·72

The measured density (by flotation in thallous malonate/thallous formate solution) is 3·86 g/cm^3

$$\mu = 3\cdot86 \times 155\cdot72 = 601\cdot02 \text{ cm}^{-1}$$
$$2/\mu = 0\cdot03 \text{ mm}$$

Thus, the optimum crystal size for cis-Pt(NH$_3$)$_2$Cl$_2$ is 0·03 mm.

Bearing in mind this restriction on crystal size, the sample can now be examined under the microscope, and suitable crystals selected. The easiest way of doing this is to use a binocular microscope which has a scale engraved on one of the eyepieces. This can be calibrated against, for example, a millimetre square on a sheet of graph paper. Several crystals are selected and placed on a microscope slide in preparation for examination under a polarising microscope. When selecting crystals it should be borne in mind that, at a later date, it may be necessary to apply an absorption correction to the diffraction data, and this is easier and less costly in computing time if the crystal shape can be approximated to a sphere or a cylinder. Crystals with re-entrant angles should not be selected as the application of an absorption correction in this case is difficult, and such crystals may also prove to be twins.

If it is decided that the sample will need to be recrystallised in order to obtain suitable crystals, the following techniques may be useful.

Crystals often stick to the container in which they are recrystallised and if they display prominent cleavage directions, attempts to dislodge them will result in the crystals splitting. Often the splits are so fine they are not observed until later when X-ray photographs are being taken. To overcome this problem the recrystallisation can be done on non-stick Teflon (polytetrafluoroethylene). Alternatively, a sheet of polythene can be used and the crystals dislodged by stretching it.

Cleavage directions also have their advantages, and large crystals can be cut to the required size under a microscope using a scalpel or razor blade which has been broken to give a pointed cutting edge, i.e. broken along a diagonal. If this is not possible a crystal can be reduced in size by pushing it around in a drop of solvent on a microscope slide until it is the required size, when it is simply pushed out of the drop of solvent. However, the evaporating solvent can sometimes loosen the lens of the microscope, so this procedure should be used with care.

Spherical crystals may be produced by using a grinder that is operated by an air jet (Bond, W. L. *Rev. Sci. Inst.*, **22,** 344 (1951). In effect the crystal is blown round a cylinder lined with carborundum powder until it is ground down to the required size of sphere. This procedure may crack fragile crystals.

When it is desirable to cut a crystal on a microscope slide, a film of vaseline on the slide prevents the crystal jumping away as it is cut. Vaseline is also useful for coating crystals to prevent oxidation by the air. It is essential when a crystal is finally selected that no satellite pieces are attached to it as these will also diffract X-rays and may interfere with the intensity data. Such satellites can be removed by pushing the crystal around in a blob of vaseline, when the viscous drag will remove them. The excess vaseline can then be removed by pushing the crystal around on the microscope slide. The crystal can be manipulated during these operations by means of a glass fibre (such as one taken from a piece of glass wool or one drawn out from a glass rod) held in a piece of plasticine.

OPTICAL EXAMINATION AND THE POLARISING MICROSCOPE

The subject of optical crystallography is well developed and standard procedures are described in several textbooks to enable the precise determination of a crystal's optical properties. (See, for instance, *Crystals and the Polarising Microscope* by N. H. Hartshorne and A. Stuart, or *Optical Crystallography* by E. A. Wahlstrom.)

Although a full examination of a crystal's optical properties can be useful to someone concerned with X-ray structure analysis, the subject of this section will restrict itself to describing how use may be made of the polarising microscope to determine the orientation of the crystallographic axes in the crystal. This knowledge facilitates the orientation of the crystal on the X-ray camera.

It will be necessary first of all to consider the nature of light itself, the way in which a polarising microscope operates, and the way in which the behaviour of light may be described as it passes through crystals of the different crystal systems.

Light may be described in terms of both the quantum theory and the electromagnetic theory, but it is the latter which is more useful to the optical crystallographer. A simple electromagnetic wave consists of an electric vector oscillating in phase with, but at right angles to, a magnetic vector.

Consider Figure 2.1(a). As the point P rotates with uniform circular motion the point P^1 executes simple harmonic motion

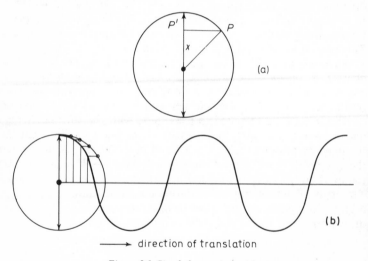

direction of translation

Figure 2.1. Simple harmonic motion

along the diameter. A sinusoidal wave may be produced by combining translation with the displacement, x, of P^1, Figure 2.1(b). Light may then be described in terms of two such waves, one representing the magnetic vector i.e.

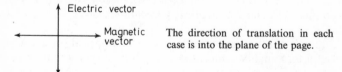

The direction of translation in each case is into the plane of the page.

If we restrict our attention to the electric vector, then ordinary light can be seen to consist of rays whose electric vectors vibrate in all possible directions at right angles to the direction in which the rays are travelling. The possible directions of vibration may be restricted so that a particular beam of light will vibrate perpendicularly to its direction of propagation but in one plane only. Light vibrating at

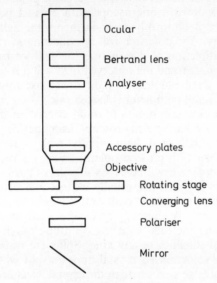

Ocular

Bertrand lens

Analyser

Accessory plates

Objective

Rotating stage

Converging lens

Polariser

Mirror

Figure 2.2. Polarising microscope

right angles to the direction of propogation in one plane only, is *plane polarised* light. The polarising microscope enables us to examine crystals in plane polarised light, and to observe the effect the crystal has on the plane of polarisation.

Figure 2.2 shows the construction of a typical polarising microscope. The analyser and polariser each have the ability to polarise light which passes through them. Modern instruments use polaroids but originally Nicol prisms made from clear calcite were used. Light which enters an optically anisotropic crystal, such as calcite, in any direction other than that of the optic axis (which is described later) is resolved into two components vibrating in mutually perpendicular planes (i.e. double refraction occurs). The Nicol prism is so constructed that only one of these components, the extraordinary ray, passes through the prism. The other, the ordinary ray is totally internally reflected. The extraordinary ray, then, emerges as plane polarised light, with its electric vector vibrating in a plane parallel to the lens of the microscope. The polariser is usually left in position during an examination of a crystal, and is so oriented that the vibration direction of the emerging light lies fore-and-aft across the microscope. The analyser usually has its polarisation direction at right angles to that of the polariser, and when a crystal is observed with both in place, the crystal is said to be observed between crossed polars.

If one looks down a microscope with crossed polars the field that is observed is uniformly dark. In theory the light entering the microscope from the polariser will be vibrating at right angles to the vibration direction of the analyser; it will not be able to pass through the analyser and the observed field will not be illuminated. In practice a small amount of light does pass through and the observed field is not completely blacked out.

A typical optical examination could consist of examining the crystals under a polarising microscope using neither polariser nor analyser; using polarised light, firstly with only the polariser inserted and then with crossed polars; and using convergent light with the convergent lens and crossed polars. Various accessories may also be used which are usually wedges and plates of various kinds which allow controlled path differences to be introduced into the optical system. A preliminary examination may be made to look for crystal defects such as splitting, obvious twinning, occlusions, and satellites of any kind. Splits are usually obvious if they exist, but sometimes a crystal appears split when in fact the split is a crystal edge seen through the crystal. Occlusions appear as small bubbles in the crystal, and obvious twins and satellites can be seen as separate entities attached to the crystal under examination.

Several apparently suitable crystals are chosen, placed on a microscope slide and examined using crossed polars[13]. By studying several crystals at once a variety of aspects of the crystals are presented and a detailed examination may be made more easily.

Crystals which are not opaque can be either isotropic or aniso-tropic. Crystals of the cubic system are isotropic and when the microscope stage is rotated between crossed polars these crystals remain uniformly dark, i.e. the light passing through them is extinguished by the analyser. Anisotropic crystals, on the other hand, are only extinguished in certain special orientations. In order to see how use can be made of this property it is necessary to look more closely at the behaviour of light rays as they pass through an anisotropic crystal.

There are two types of anisotropic crystal, those which are uniaxial and those which are biaxial. In the uniaxial case, light entering the crystal is split into an ordinary and an extraordinary ray, whereas in the biaxial case the light splits into two extraordinary rays. The behaviour of these rays on passing through a crystal is shown by means of a geometric construction known as the indicatrix. This is an ellipsoid whose dimension in any direction from the centre is proportional to the refractive index of the light vibrating along that direction. The uniaxial indicatrix has two of its principal axes equal and the third can be larger or smaller than these; the biaxial indicatrix is a triaxial ellipsoid having its three mutually perpendicular semi-axes equal to the three principal indices of refraction of a biaxial crystal.

The indicatrix can be related to the crystallographic axes by choosing a common origin, and in this way a relationship can be deduced between the positions at which a crystal extinguishes between crossed polars and the crystallographic axes of the crystal. An optic axis is a direction in the indicatrix in which there is no double refraction. Uniaxial crystals can belong to the trigonal, tetragonal, or hexagonal crystal systems and the optic axis coincides with the symmetry axis of the crystal, the c-axis*. Biaxial crystals have two optic axes and can belong to the orthorhombic, monoclinic, or triclinic crystal systems, and the relationship of the indicatrix to the crystallographic axes will differ in each case. In orthorhombic crystals the principal axes of the indicatrix coincide with the crystallographic axes, Monoclinic crystals can have any one of the indicatrix axes coincident with the unique b-crystallographic axis; the other two indicatrix axes will then lie in the ac plane. In triclinic crystals the indicatrix can have any orientation with respect to the crystallographic axes.

If a crystal is extinguished, i.e. displays extinction when viewed in parallel light between crossed polars, then the vibration directions of the polars coincide with a principal vibration direction in the crystal (see Figure 2.3). It must, however, be borne in mind that the principal vibration directions in the crystal need have no relation to the presentation of the edges of the crystal as seen between crossed polars. This is particularly true of triclinic crystals. Two types of extinction can arise. The first is seen with both cubic crystals when viewed from any direction, and uniaxial and biaxial crystals when viewed down an optic axis. In these cases the crystal remains extinguished throughout the rotation of the microscope stage. In the second type of extinction the crystal will extinguish in four positions during the rotation of the microscope stage, each

*Crystals which belong to the trigonal system may have their unit cell based on (a) a set of rhombohedral axes or (b) a set of hexagonal axes (see *Table 1.1*). In the former case the symmetry axis lies along a body diagonal, and in the latter case the symmetry axis coincides with the c-axis.

position being separated by an angle of 90° from the adjacent extinction positions. This type of extinction occurs with every aspect of an anisotropic crystal other than those already mentioned, and the vibration directions in the crystal are the axes of the indicatrix section that is parallel to the crystal plate.

How may the knowledge of the extinction directions in a crystal be put to use?

If extinction is observed in a crystal during the rotation of the microscope stage, even when the crystal is viewed from several different directions, then in all probability it is a cubic crystal. If a crystal gives persistent extinction only when viewed from one direction it is uniaxial, and if it gives persistent extinction when viewed from two different directions it is biaxial. It is possible to

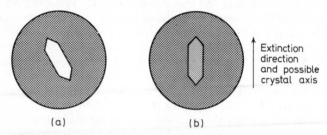

(a)   (b)

Figure 2.3. An extinction direction in a crystal. (a) No extinction. Background dark due to crossed polars; crystal illuminated. (b) The extinction direction. Background and crystal dark; the vibration directions of the crossed polars coincide with a principal vibration direction in the crystal

differentiate between uniaxial and biaxial crystals by examining their interference figures seen between crossed polars, and using convergent light from the converging lens shown in the microscope diagram (Figure 2.2). The figures may be observed by either removing the microscope eyepiece, or inserting a Bertrand lens into the microscope. In addition an objective with a high resolving power is used. The Bertrand lens combines with the eyepiece to form a low power microscope focussed on the upper focal plane of the objective.

Uniaxial and biaxial crystals when viewed from a direction other than the optic axial direction give the four extinction positions during the rotation of the microscope stage that are typical of anisotropic crystals.

Extinction can be straight (parallel to a crystal edge), symmetrical (bisecting two edges), or oblique (lying at an angle not parallel to a crystal edge, and not bisecting two edges). Uniaxial crystals can show straight or symmetrical extinction. The extinction shown by biaxial crystals depends upon the crystal system to which they belong. In the orthorhombic system general faces show oblique extinction, but faces belonging to zones whose edges are crystallographic axes (i.e. sections having one vibration direction corresponding to a principal refractive index) can show straight or symmetrical extinction as in the uniaxial case. In the monoclinic system straight extinction will occur in the [010] direction i.e. when faces parallel to the unique b-axes of the crystal are examined. Any other faces will give oblique extinction. If the crystal belongs to the triclinic system oblique extinction occurs although in certain cases straight extinction may be seen.

The purpose of the examination of the crystal sample between crossed polars is therefore to find a crystal so oriented that extinction is observed parallel to a prominent crystal edge, in the hope that this edge or a direction at right angles to it will be a crystallographic symmetry axis. (This, of course, does not apply to those crystals belonging to the triclinic system except accidentally.) The symmetry of the X-ray photographs will show which of the possible choices is the symmetry axis.

It can be seen that in many cases a careful study of both the crystal morphology and the crystal extinction directions can enable the crystal to be mounted on the X-ray camera, so that it rotates about a crystallographic axis.

During the examination of anisotropic crystals between crossed polars, crystal twinning will show up as differences of illumination in different areas of the specimen, although it may be necessary to examine the specimen from different directions before this is seen. Twinned crystals should only be used as a last resort, as the unravelling of the X-ray diffraction patterns is very difficult.

Before proceeding further, the dimensions of the crystal should be measured with the calibrated eyepiece of the microscope. These will be needed later when an absorption correction is applied to the intensity data.

MOUNTING THE CRYSTAL ON A GONIOMETER HEAD

The selected crystal may now be mounted on the goniometer head in preparation for the collection of X-ray photographic data.

A goniometer head, or set of arcs, consists of a central pin below which are two adjustable slides in the shape of arcs, which are at right angles to each other and graduated in degrees. When both arcs are in the zero position the pin will be vertical. The pin may be tilted from the vertical by a movement of either of the arcs. Beneath the arcs are two further slides which are also at right angles to each other. Lateral movement of these slides is used to ensure that the crystal when attached to the pin will lie on the axis of rotation of the goniometer head, i.e. the crystal will lie on the axis of rotation of the Weissenberg camera, as the goniometer head is attached to the camera, and all adjustments to the goniometer slides are made when the goniometer is on the camera.

The arcs have a common centre about which they move and the ideal situation exists when the crystal lies on this centre, as any corrections to the crystal setting by means of the arcs will not then

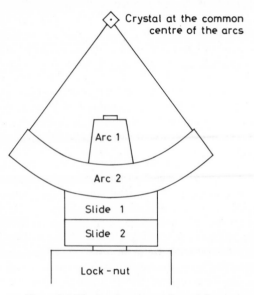

Figure 2.4. The rotation centre of a set of arcs.
Arc 1 is at right angles to arc 2

involve re-adjusting the other slides to re-centre the crystal on the camera axis of rotation. Figure 2.4 shows the centre of rotation of the arcs.

The base of the goniometer head has a lock-nut that attaches it to the rotating spindle of the Weissenberg camera.

It is usual to attach the crystal to the pin of the goniometer head by means of a fine glass fibre which should be rigid. A suitable fibre can be drawn out from a piece of glass tubing (a glass wool fibre is a little too fine). The fibre is attached to the pin by an adhesive such as nail varnish, and its length adjusted so that when the crystal is attached to the tip it will lie on the centre of rotation of the arcs.

The fibre tip is then finely coated with adhesive and brought up under the microscope to the crystal, which will adhere to the fibre in the required orientation—usually so that the extinction direction is parallel to the axis of rotation of the goniometer head. Before the adhesive sets the crystal can be further orientated by gently pushing it with a fine glass fibre.

Thought should now be given to the reason for mounting the crystal. If it is for a preliminary examination, the determination of unit cell parameters and space group, then the one crystal can be used for all the determinations as long as its position on the fibre can be altered to allow it to rotate about each of its three crystallo-graphic axes in turn. (This is not always possible; for example if the crystals are fine fibres it may be difficult to mount them about an axis other than the fibre axis. In this case Weissenberg and preces-sion photographs may be used to obtain information about cell parameters and space groups.) In the above case the crystal can be attached to the fibre by an adhesive such as vaseline or silicone grease. If, however, it is intended to use the crystal on a Weissen-berg camera for the collection of X-ray intensity data, then it is advisable if possible to use three crystals, each one mounted so as to rotate about a different crystallographic axis. In this case an adhesive such as shellac or nail varnish can be used. If the intensity data are to be collected on a diffractometer the crystal must be held extremely rigidly and usually a quartz fibre is used with Araldite as an adhesive both for the fibre and the crystal. In fact goniometer heads are not usually rigid enough for diffractometer work and on 4-circle diffractometers a steel spike is often used instead.

When the crystal is to be cooled by a liquid nitrogen cooling system during the data collection it is not advisable to use silicone grease as an adhesive if it will be required to reorient the crystal on the fibre, as the grease can set solid and remain so even when warmed to room temperature again. For low temperature work the crystal can be sealed into a Lindemann capillary tube of a suitable diameter, as this protects the crystal from the snow which may otherwise form on it. Lindemann capillary tubes are also useful if the crystal needs to be kept in an inert atmosphere, and for crystals that need to be kept free from water vapour a desiccant can be put at the base of the tube. An additional application of Lindemann tubes is their use with liquid samples that need to be crystallised by cooling before X-ray data is collected.

A convenient way of mounting a crystal in a capillary tube is to attach the tube to the goniometer head with an adhesive such as nail varnish (plasticine can also be used). The crystal is attached near the open end of the tube by means of a smear of adhesive which has been inserted into the tube on the end of a fine fibre. The tube is then sealed with a spot of nail varnish. The insertion of the crystal into the tube is carried out under a binocular microscope and the crystal can be conveniently transferred from slide to tube on the end of a glass-wool fibre which has been lightly coated with a fine

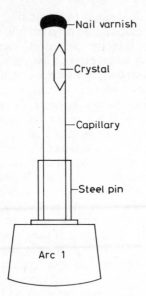

Figure 2.5. A crystal in a capillary tube on a set of arcs. The needle axis of the crystal is parallel to the rotation axis of the goniometer head

film of vaseline as an adhesive. Figure 2.5 shows a crystal mounted in a capillary tube.

X-RAY EXAMINATION

The point has now been reached where X-ray data can be collected from the crystal. If the intensity data are to be collected using a diffractometer, the unit cell dimensions and space group of the sample are usually determined initially by film methods. All the necessary photographs can be taken using a Weissenberg camera, and the information that is obtained can be confirmed or added to by the use of a precession camera. While a rotation camera is useful, it is not essential as the Weissenberg camera can be used as a rotation camera with the added advantage that once the rotation photographs have been taken, the crystal is already aligned for taking Weissenberg photographs.

From oscillation or rotation photographs the length of the axis of rotation can be obtained. If a Weissenberg photograph is taken, for example, with the crystal rotating about the a-axis, then the lengths of the reciprocal axes b^* and c^* and the angle between them α^* can be found. The number of molecules per unit cell of the sample can be determined from a knowledge of its density, its molecular weight, and its cell volume. The unit cell volume can be obtained from a rotation or oscillation photograph together with a Weissenberg photograph taken with the crystal rotating about the same axis. In this way a check can be made on the chemical composition of the sample. Cell parameters are obtained from measurements made on oscillation (or rotation) and Weissenberg photographs and space groups may be deduced from absences noted in the diffraction patterns of Weissenberg photographs. A procedure that may be followed in acquiring the above information is described below.

SETTING UP THE CRYSTAL ON A WEISSENBERG CAMERA

In order to obtain photographs that have any value, the crystal must be accurately aligned so that one of its crystallographic axes coincides with the rotation axis of the camera, which is usually horizontal. See Figure 2.6. To help in this aim the crystal has been attached to its fibre so that an extinction direction coincides with the camera rotation axis. Further fine adjustments will be necessary

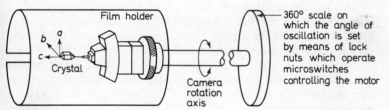

Figure 2.6. A crystal on a Weissenberg camera. The X-ray beam reaches the crystal through the slit in the camera holder, and is at right angles to the rotation axis

to make perfect this alignment, and the magnitude of these adjustments is obtained from photographic data.

The goniometer head is mounted on the spindle of the Weissenberg camera, and the position of the spindle is adjusted so that the crystal will lie in the path of the X-ray beam. By means of a telescope attached to the camera the crystal is viewed while the spindle is rotated by hand. During this rotation the crystal should remain centred on the axis of rotation. If it does not then adjustments are made to the goniometer head slides in the following way:

1. The position of the crystal is noted with respect to the telescope cross wires when one of the goniometer head slides is vertical.
2. The crystal is rotated through 180° and its position again observed.
3. The vertical slide is adjusted so that the crystal lies halfway between these two positions.
4. The procedure is then repeated with the other slide vertical.
5. The crystal is rotated through 360° and checked that it is now centred on the camera rotation axis. If it is not centred the above procedure 1, 2, 3, 4 is repeated.

It may be necessary to make some adjustment to the goniometer head arcs at this point, for example, if the extinction direction in the crystal lies along a needle axis but the needle when viewed through the telescope is not lying exactly along the rotation axis of the camera. As long as the crystal lies at the centre of rotation of the arcs themselves, then adjustments can be made to the arcs alone. However, if the crystal does not lie on this centre, any adjustment to the arcs must be followed by a re-centring of the crystal using the goniometer head slides.

An initial setting photograph may now be taken; the telescope is racked back, the beam-stop is inserted in the instrument, one of the arcs of the goniometer head is set in a vertical position, the camera spindle is geared to carry out a 15° oscillation. In the dark room a mark is made on the film so that its orientation in the film holder is known, the film is wrapped in black paper and inserted in the film holder which is then locked on to the camera so that the crystal lies at the centre of the camera's cylindrical axis; the X-rays are switched on and the crystal is made to oscillate continually through 15° by the camera motor. After about half an hour the film can be taken off and developed. It is good practice always to mark a corner of the film, i.e. initial it in pencil, and insert the film into the film holder so that the initialled corner lies in the same place each time. The pencil mark is then used as a reference point when corrections are made to the goniometer arcs from measurements made on the film after it has been developed.

If the crystal was perfectly set about a crystallographic axis the photograph will be similar to Figure 2.7(a), i.e. there will be parallel layer lines of spots symmetrically distributed about the equatorial zero layer line. How this sort of diffraction pattern arises is explained later.

It is more likely, however, that further adjustments to the crystal setting will need to be made by means of the arcs. Figures 2.7(b) to (e) show the sort of patterns obtained using mis-set crystals.

THE DETERMINATION OF CORRECTIONS TO THE CRYSTAL SETTING

The above procedure for taking photographs of X-ray diffraction patterns is usually modified a little to make more easy the determination of the corrections to be applied to the arcs. There are many ways of taking setting photographs and two methods are described below. The first, is a rule of thumb method which works well when large corrections are necessary, and the second is a more precise calculation more applicable when small corrections are all that is needed.

Rule of Thumb Method

An oscillation photograph is taken as described above, and then a further exposure of about 10 minutes is made with one of the camera screens in place so that a shadow of the screen falls along the zero layer line of the oscillation photograph. This gives a straight equatorial line from which measurements of the degree of mis-setting can be made, e.g. see Figure 2.7(a). It is better to use a line such as this from which to measure the angular mis-setting of the zero layer line than using the film edge, as there is no guarantee that the film is correctly aligned in the film holder. Figures 2.7(a)– 2.7(g) illustrate the method of obtaining the angular correction to be applied to the arcs from the oscillation photograph. In each case the arcs are shown superimposed on the film in exactly the position they occupied when the film was taken. The film has been opened out so that it is now flat, not cylindrical.

Figure 2.7(a) shows the ideal situation where the zero layer line is parallel with the shadow of the X-ray screen.

To decide how to apply corrections to the arcs, it is convenient to think of the crystal as a cylinder and the zero layer line as the cylinder base line. If a picture is obtained as in Figures 2.7(b) or 2.7(c), then the cylinder has to be tilted, as shown, by $\theta°$ to make its base a horizontal line.

If, on the other hand, the picture looks like Figure 2.7(d), then the cylinder (crystal) has to be moved towards the observer by $\theta°$ to again make the cylinder base a horizontal line. If the picture appears

44

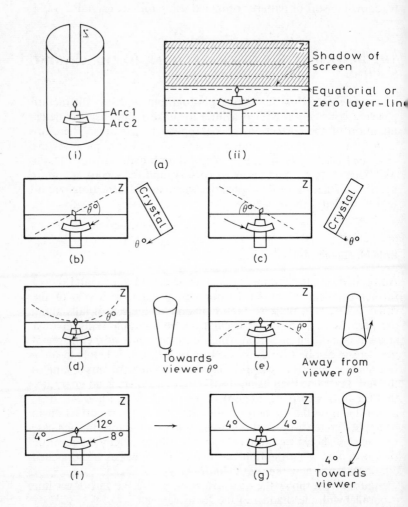

Figure 2.7. Oscillation photographs illustrating the rule-of-thumb method. The drawings are symbolic and do not show any symmetry in the diffraction patterns. (a) Shows the relationship between the arcs and the film (i) while in the camera, and (ii) after development. The shadow of the screen is shown; this can be used as a reference line for making measurements of the degree of mis-setting of the crystal. (b)–(g) Show the corrections that need to be applied to the arcs to correct the mis-setting of the crystal by the rule-of-thumb method applied to the zero layer-line

as Figure 2.7(e) then the cylinder has to be moved away from the observer by $\theta°$ to make the cylinder base a horizontal line.

Usually an intermediate position exists as in Figure 2.7(f). In this case corrections to both arcs are necessary. Firstly arc 2 is moved 8° as shown, which would result in a picture like Figure 2.7(g). A 4° correction is then made to arc 1 in the usual manner which should result in the crystal being correctly set.

Double Oscillation Method

Many references can be found in the literature to ways of setting crystals accurately for X-ray photographs[14, 15, 16, 17]. One method described here makes use of a double oscillation photograph. The goniometer arcs are arranged so that one is parallel to the X-ray beam and one is perpendicular. A 15° oscillation photograph is taken about this position. The arcs are then rotated through 180° and the procedure is repeated for a longer period of time. In this way two diffraction patterns are obtained on the one film, and the equatorial base line from which the corrections may be measured is obtained by drawing a horizontal line across the film with the two zero layer lines used as markers.

Kratby and Krebs[17], and Hendershot[14], have shown that the errors on each goniometer arc can be calculated from the shape of the zero layer line on an oscillation photograph. Hendershot has shown that for an angular error i_1 in the arc perpendicular to the beam

$$d_1 = R \sin 2\theta \sin i_1$$

and for an angular error i_2 in the arc parallel to the beam

$$d_2 = R(1 - \cos 2\theta) \sin i_2$$

where d_1 and d_2 are the distances of displaced points from the ideal position of the zero layer line for a Bragg angle θ. R is the camera radius. When $\theta = 90°$, d_2 is a maximum and d_1 zero. The displacement at $\theta = 90°$ is due solely to an error in the arc set parallel to the beam, while the displacement at $\theta = 45°$ is due on one side of the film to the sum and on the other to the difference of the errors in the two arcs. By measuring the displacements $(d_1 + d_2)$ and $(d_1 - d_2)$ at $\theta = 45°$, d_1 and d_2 are obtained and i_1 and i_2 derived from the above equation. A further value of i_2 is obtained by measurement of the displacement $2d_2$ close to $\theta = 90°$.

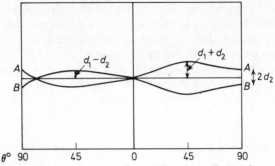

Figure 2.8. Arc corrections obtained from an oscillation photo-graph. AA and BB are the two zero layer lines. For a camera of radius 2·865 cm, $2d_1$ and $2d_2$ in millimetres are equal to the errors in degrees in the respective arcs i_1 and i_2

Figure 2.8 shows how use is made of a double oscillation photo-graph to obtain the corrections to apply to the goniometer arcs.

THE DETERMINATION OF UNIT CELL DIMENSIONS FROM OSCILLATION OR ROTATION PHOTOGRAPHS

A rotation or oscillation photograph contains information that leads to a knowledge of the length of the crystallographic axis about which the crystal was rotated while the photograph was taken. How this arises can be explained as follows.

Diffraction at a set of planes can be represented as in Figure 2.9.

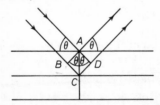

Figure 2.9. Diffraction at a set of planes

For reinforcement to occur there must be a whole number of wave-lengths difference in the path lengths of rays scattered by successive planes.

In Figure 2.9 this means that BCD must be a whole number of wavelengths.

Now $BC = CD = d \sin \theta$

$BCD = 2d \sin \theta$

For reinforcement $n\lambda = 2d \sin \theta$ where n is a whole number. This is

Bragg's law, and every time a 'reflection' occurs, this equation has been satisfied by the geometry of the crystal and the incident X-ray beam. If we now consider any oscillation photograph, the zero layer line is seen to arise from reflections of the incident X-ray beam by sets of planes parallel to the axis of rotation of the crystal. Figure

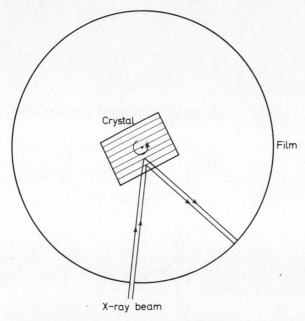

Figure 2.10. Reflecting planes in a crystal. Diffraction by a set of planes parallel to the crystal rotation axis

2.10 shows a plan view of such a diffracted beam or reflection occurring from one such set of planes.

As the crystal rotates about the *a*-axis, the family of 0*kl* planes will produce the zero layer reflections as successive sets of planes are brought into positions that satisfy the Bragg equation.

Now while these reflections are occurring, other planes not parallel to the axis of rotation will also be satisfying the Bragg equation. This results in the production of cones of reflections by general *hkl* planes as shown in Figure 2.11. Each cone has a different value of *n* in the Bragg equation and is produced by a family of planes *nkl*. The film in the X-ray camera occupies the position of the cylinder enclosing these cones shown in Figure 2.11, and each cone will therefore appear as a straight line of diffraction spots on the film.

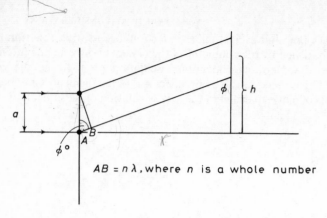

$AB = n\lambda$, where n is a whole number

Figure 2.11. Diffraction cones produced by a rotating crystal

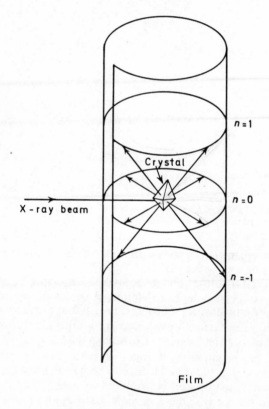

Figure 2.12. Layer-line spacing and the rotation axis repeat unit

A spot appears on the cone n whenever a set of planes nkl satisfies the Bragg equation.

CALCULATION OF THE LENGTH OF THE ROTATION-AXIS[18]

Figure 2.12 shows a diffracted ray on a cone of semi-vertical angle meeting the film at a height h above the zero layer line. From the figure it can be seen that:

$$n\lambda/a = \cos \phi,$$
$$a = n\lambda/\cos \phi,$$
$$\tan \phi = r/h$$

where r = camera radius.

By measuring the height (h) of a layer line from the zero layer line on the film and knowing the camera radius (r), the length of the a-axis can be obtained. It is usual to use the higher layer lines when determining cell lengths by this method and to measure $2h$, the distance between similar layer lines above and below the equator. If the crystal is rotated about each of its axes a, b, c, in turn then all the axial lengths can be found, one from each of the oscillation or rotation photographs that are taken. In fact, rotation photographs are preferable, as any mis-setting of the crystal will manifest itself as an uneven layer line. Often, however, only rough values are required from oscillation photographs, and accurate cell dimensions are obtained from calibrated Weissenberg photographs by methods which are described later.

WEISSENBERG PHOTOGRAPHS

Weissenberg photographs differ from oscillation photographs in that data are collected from only one layer line at a time. Figure 2.13 shows how this is done. Cylindrical screens are arranged coaxially around the crystal rotation axis so that only diffracted rays lying on one layer line, i.e. one cone of diffracted rays, are allowed to fall on the photographic film.

The photographic data are collected while the crystal rotates through an angle of about 180° and at the same time the film holder is given a translational movement, oscillating along the rotational axis of the instrument. Throughout this procedure the screens remain stationary so that only the one layer line of data is collected. The crystal rotation and film-holder translation are usually geared so that 1° of rotation is equivalent to 2 mm of traverse. Figure 2.14 shows an idealised Weissenberg camera.

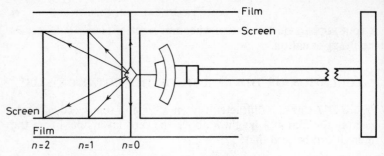

Figure 2.13. Collection of intensity data from one layer line only, n = 0, by the use of screens to prevent other layers registering on the film. Only positive values of n are considered for clarity

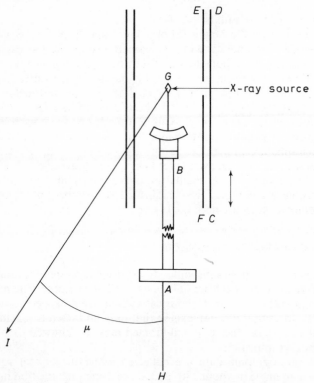

Figure 2.14. Representation of an idealised Weissenberg camera. 1. The crystal can oscillate through any pre-set angle about the axis AB. 2. The film-holder CD can oscillate through any preset distance along CD. 3. The screens EF remain stationary during data collection, but can be translated along EF to allow different diffracted cones to fall on the film. 4. The axis of the above instrument can be swung through an angle μ, i.e. GH goes to GI, in order to satisfy the requirements for data collection from higher layers

To understand the data produced by this mechanism, it is necessary to consider the reciprocal lattice of the crystal. It was shown earlier that each cone of diffracted rays from a crystal is produced by a family of planes, e.g. *nkl* if the crystal rotates about the *a*-axis, where *n* is a whole number. Each spot on a layer line is produced by a set of planes in the crystal such as the 001 set, and this is represented by a point on the reciprocal lattice. Each layer line represents the projection of a two-dimensional net of the reciprocal lattice on to the film. The indices of this net being *nkl*. The zero layer line comes from the 0*kl* layer of the reciprocal lattice. Now the operation of the Weissenberg instrument is such that the layer line of the oscillation photograph is drawn out on a

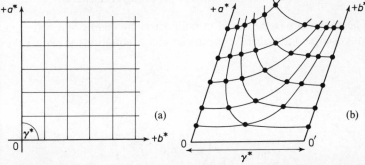

Figure 2.15. The reciprocal lattice and the Weissenberg photograph. (a) A quadrant of the reciprocal lattice. (b) An idealised representation of a quadrant of the reciprocal lattice as seen on a Weissenberg photograph. The origin, 0, of the +b* axis in (a) is translated to 0′ in (b). 00′ in millimetres equals half the angle γ* in degrees, when the camera has a rotation/translation ratio of 1° to 2 mm. 0a* and 0′b* are parallel

Weissenberg photograph to give a distorted picture of a two-dimensional layer of the reciprocal lattice, each point on this lattice being a reflection produced by a particular set of planes in real space. Figure 2.15 shows the distortion that must be applied to a layer of the reciprocal lattice to produce the pattern found on a Weissenberg photograph.

From the zero layer Weissenberg photograph it is possible to obtain the values of two reciprocal cell lengths and the angle between them.

DETERMINATION OF RECIPROCAL CELL LENGTHS FROM ZERO LAYER WEISSENBERG PHOTOGRAPHS

According to reciprocal lattice theory[19] an X-ray reflection will occur whenever a rotating reciprocal lattice point touches the sphere of reflection. Figure 2.16 shows a plan view of the zero layer of the reciprocal lattice rotating at B. A point of the lattice cuts the sphere of reflection at P. The outer circle represents the film and the diffracted ray from the crystal meets the film at R. $P\hat{O}B = 2\theta$.

The vertical* distance (x) of a spot on a Weissenberg film from the equator of the film is related to the radius of the film (r) and the Bragg angle θ by the expression[20]: $\dfrac{x}{\pi r} = \dfrac{2\theta}{180}$

r is usually chosen such that $\pi r = 90°$; x then equals θ. By Bragg's law: $\lambda = 2d \sin \theta$

$$\frac{\lambda}{d} = 2 \sin \theta = Z^* \text{ the reciprocal cell dimension}$$

The distance x is measured in millimetres by means of a travelling microscope and substitution in the above equations gives a value

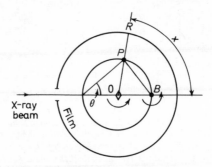

Figure 2.16. The production of a diffracted
ray by the zero layer of the reciprocal lattice

of the reciprocal lattice repeat distance. Usually spots of high θ value are used.

THE DETERMINATION OF RECIPROCAL CELL ANGLES

1. The reciprocal axes can be drawn on the film with a needle and the distance between them at the points where they cut the equator of the film can be measured with a travelling microscope. If, as is usual, the crystal rotation is related to the camera traverse so that 1° of rotation equals 2 mm of traverse, then twice the distance between the axes in millimetres gives the angle between the axes.

*The vertical distance of a spot from the equator is the shortest distance i.e. 'vertically' down the film.

2. The length of a diagonal of the reciprocal cell can be determined by measuring the vertical distance of, for example, the 606 spot from the equator and calculating its length as for an axial length. Substitution in the cosine equation will then give the angle. Figure 2.17 illustrates this. *BD* is calculated, $b^2 = c^2 + a^2 - 2ac \cos(180 - \beta)$ hence β is obtained.

In practice Weissenberg films are usually calibrated by means of

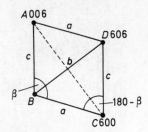

Figure 2.17. Calculation of an inter-axial angle

a powder photograph of known *d*-spacings, so that more accurate cell dimensions may be obtained.

CALIBRATION OF WEISSENBERG FILMS

A Weissenberg photograph is taken of the particular zero layer of the reciprocal lattice that is to be examined. The set of arcs holding the crystal is then replaced by a new set of arcs on which is mounted an aluminium wire. The wire is centred so that it rotates along the axis of the instrument. The film-holder is disengaged from the traverse drive and moved to a position where the screens, which are set a few millimetres apart, will allow a powder diffraction pattern from the wire to fall on the extreme edge of the film. A rotation photograph of the wire is thus superimposed down one side of the film on which the zero layer data have been collected. After exposure for one to two hours the procedure is repeated so that a similar rotation photograph is superimposed down the other edge of the film. The resulting photograph will look something like Figure 2.18, which shows a zero layer Weissenberg photograph of a copper glycyl-L-tryptophane complex with a gold powder pattern superimposed down each side. The positions of the powder lines with respect to the equator of the film can be calculated simply from the known *d* spacings for the material of the wire and the wavelengths of the radiation. Table 2.1 shows the values of *d* spacings for aluminium[21] and the wavelength of CuK_{α_1} and CuK_{α_2} radiation.

A table of wavelengths of radiation obtained from different target materials is given in *International Tables for X-ray Crystallography—Vol. III.*

Use can be made of the superimposed powder patterns in the following ways. Lines parallel to the equator are drawn on the film

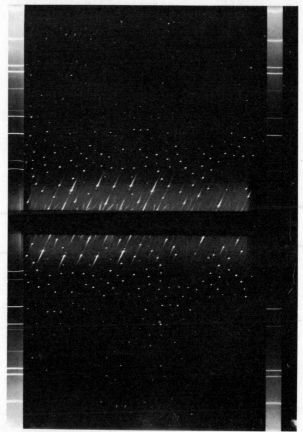

Figure 2.18. A Weissenberg photograph calibrated with gold wire powder lines

Table 2.1 COPPER WAVELENGTHS AND ALUMINIUM d SPACINGS

$$\left. \begin{array}{l} \text{Cu}K_{\alpha_1} = 1{\cdot}540\,51\,\text{Å} \\ \text{Cu}K_{\alpha_2} = 1{\cdot}544\,33\,\text{Å} \end{array} \right\} \quad \text{Mean} = 1{\cdot}542\,42\,\text{Å}$$

hkl	111	200	220	311	222	400	331	420	422
d Å	2·338	2·024	1·431	1·221	1·169	1·0124	0·9289	0·9055	0·8266

with a needle joining up the two powder patterns. Then one of the following procedures can be followed:

1. The vertical distance of a spot on the Weissenberg photograph from the equator of the film can be obtained by measuring its distance from the nearest needle-drawn line and adding this to the known value of that line. Alternatively, the vertical distance of spots or arcs from the equator may be obtained by measuring the distance between similar arcs or spots above and below the equator and halving the value obtained.
2. The distances of high θ value needle-lines from the equator are measured and a correction factor calculated for the film based on the known values of these lines for the powder pattern. This factor is then applied to the measured vertical distance of the spots from the equator.

In this way, allowance is made for errors due to film shrinkage, and at the same time base lines are provided from which to make measurements so that any eccentricity of the film in the film holder does not interfere with the calculation of cell dimensions.

A 'least squares' procedure may be used to obtain the best set of cell parameters from measured values of several spots on the same reciprocal axis. More accurate cell parameters are obtained from measurements made at higher θ values.

DETERMINATION OF MOLECULAR WEIGHT

The density of the sample can usually be determined by a flotation method fairly easily*. The volume of the unit cell can be calculated from the cell parameters. It is a simple matter then to calculate the weight of matter in the unit cell which will be the molecular weight or a simple multiple of the molecular weight of the sample.

Assuming the sample belongs to the triclinic crystal system, the length of a unit cell edge is given by:

$$a = \frac{b^* c^* \sin \alpha^*}{V^*}$$

$V^* = $ reciprocal cell volume

from which
$$V^* = \frac{b^* c^* \sin \alpha^*}{a}$$

*See *International Tables for X-ray Crystallography*, Vol. III, p. 17.

but
$$V = 1/V^*$$
$$V = a/b^*c^* \sin \alpha^*$$

where V = unit cell volume

The value of a can be obtained from an oscillation photograph taken about the a-axis, and a Weissenberg photograph taken about the same axis will give the values of b^*, c^*, and α^*. V can therefore be obtained.

Now $V(\text{cm}^3) \times \sigma(\text{g/cm}^3)$ = Molecular weight $\times n$

where n = 1 or a small whole number; the number of molecules per unit cell

and σ = the density of the sample

Therefore $M = \dfrac{V}{n} \times 0.6023 \times 10^{24}$, where 0.6023×10^{24} is Avogadro's Number. Substitution in the above expression will then give M the molecular weight[22], and confirmation of the value of n (see page 66).

HIGHER LAYER WEISSENBERG PHOTOGRAPHS (equi-inclination photographs)

The procedure for taking higher layer equi-inclination Weissenberg photographs differs from that of zero layer ones in that it is necessary to slew the Weissenberg instrument around through an angle μ so that the X-ray beam, incident on the crystal, lies along a generator of the cone of reflections produced by the reciprocal lattice layer that is being examined. The screens must also be moved sideways so that the diffracted cone will fall on the film. Figures 2.19(a) and (b) illustrate the alterations made to the instrument setting.

In the plan view, Figure 2.19(c), the circle AB is similar to the equatorial circle of diffracted rays from a zero layer of the reciprocal lattice. The reflecting sphere has a radius of unity, and $CD = \zeta/2$ therefore the radius of the circle AB will be $\sqrt{(1 - \zeta^2/4)}$ and not unity as for the zero layer[23]. ζ is the co-ordinate of the reciprocal lattice layer parallel to the rotation axis. The distance S through which the screens must be translated from the zero layer position in order that the diffracted rays from a higher layer are able to fall on the film is shown in Figure 2.19(d).

With reference to Figure 2.19(c), the general condition for equi-inclination photographs is that $\mu = -\nu = \sin^{-1} \dfrac{\zeta}{2}$. The angle ν is the complement of the semi-angle of the cone of the reflected rays, and the angle μ is the complement of the angle between the incident beam and the oscillation axis.

When $\mu = \nu = 0$ the incident beam is normal to the crystal rotation axis, i.e. the arrangement for zero layer Weissenberg photographs.

M. J. Buerger in *X-ray Crystallography* p. 294–295 (John Wiley & Sons) gives two graphs from which the μ and S values for a layer

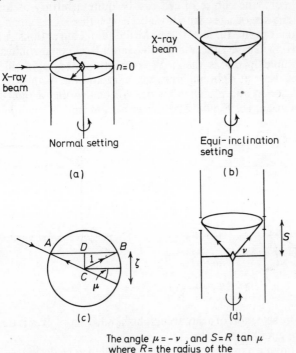

The angle $\mu = -\nu$, and $S = R \tan \mu$ where R = the radius of the screens

Figure 2.19. The geometry of the diffraction cones in the normal beam, and equi-inclination settings of a Weissenberg camera

line can be obtained if (a) the value of the ζ co-ordinate for the layer is known, or (b) the height y of the layer line from the equator of an oscillation photograph is known.

THE CHOICE OF UNIT CELL

The position has been reached where suitable crystals have been selected, and one has been mounted on a set of arcs so as to rotate about an extinction direction. By means of oscillation and Weissenberg photographs about each crystallographic axis, the lengths of the unit cell axes, a, b, c, and angles α, β, γ, can be obtained, and the unit cell volume, V, can be calculated. The question arises as to what criteria are used in choosing the unit cell of a crystal from the many unit cells available. This is the same as asking what governs the choice of reciprocal axes on the zero layer Weissenberg films.

In theory, the choice of cell can be quite arbitrary as long as a consistent set of axes are chosen, i.e. the three axes must have a common origin. In fact there are certain conventions which, if followed, will save time and trouble later. If a crystal rotates about, for example, its a-axis, then a Weissenberg photograph will show a reciprocal net at right angles to this axis, and the reciprocal axes of the net are b^* and c^*. The axes are labelled so that a right-handed set is chosen, i.e. Figure 2.20 shows a right and left-handed set of

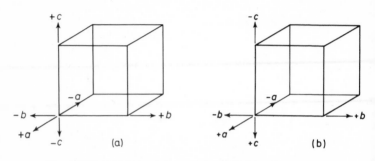

Figure 2.20(a). Right-handed set of axes. (b) Left-handed set of axes

axes. The selection of a suitable right-handed set of axes is described in more detail later.

The unique axis in a monoclinic unit cell may be designated b or c. For the purpose of the following discussion it is assumed to be the b-axis. The β angle of the monoclinic unit cell is then chosen as obtuse, and as $\beta^* = 180 - \beta$, then β^* is acute. In the triclinic cell, all three angles of the cell may be chosen obtuse. (The direction cosines of the axes with the [111] zone axis, should then all be positive.) The reduced primitive cell is usually chosen, i.e. that unit cell having cell edges which are the three shortest lattice translations. The cell edges are labelled so that $a > b > c$. $+a$, $+b$ and $+c$ can then be chosen so that α, β, and γ are all obtuse. Because of the difficulty that is sometimes encountered with triclinic crystals it is often easier simply to select α, β and γ to be as near 90° as possible, and not necessarily greater than 90°.

The relationships between the real and reciprocal angles in a triclinic cell are given in Table 1.6. The unique axis in a monoclinic crystal is labelled b or c, and in trigonal, tetragonal and hexagonal crystals it is labelled c. (Assuming the trigonal cell is described by hexagonal axes.)

In unit cells where $\alpha = \beta = \gamma = 90°$ there should be no difficulty in choosing the axes (since measuring across the film, 1 mm $\equiv 2°$, hence 4·5 cm $\equiv 90°$) which are usually self-evident.

A monoclinic crystal, rotated about the unique b-axis, will show reciprocal axes a^* and c^* separated by an angle β^* on a zero layer Weissenberg photograph; if rotated about either the a- or c-axis then a zero layer photograph would show respectively the b^* and c^* axes 90° apart, or the a^* and b^* axes 90° apart.

In the triclinic case, the choice of axes is a little more difficult.

It should be remembered that as long as a consistent set of axes has been obtained on three zero layer Weissenberg photographs, the labelling of the axes and the choice of sign of the axes can be quite arbitrary, although it is usual to follow the conventions.

If the monoclinic case is considered, the b-axis may be fixed by the symmetry of the diffraction pattern and the a-axis is the shorter of the other two. The signs of the axes are chosen so as to give a right-handed set.

A consistent set of axes will have been chosen when:

1. Rotation about the a-axis gives a zero layer Weissenberg photograph showing the b^* and c^* axes.

2. Rotation about the b-axis gives a zero layer Weissenberg photograph showing the a^* and c^* axes.

3. Rotation about the c-axis gives a zero layer Weissenberg photograph showing the a^* and b^* axes.

If this consistency is not obtained, then one of the axes of rotation has probably been wrongly chosen.

Each zero layer Weissenberg photograph will contain two reciprocal axes. Any two of the three possible zero layer photographs will have a common reciprocal axis which will be identical in both spacings and relative intensities of the spots. Figure 2.18 shows a typical Weissenberg photograph.

A monoclinic crystal rotated about a unique $+b$-axis will show a zero layer Weissenberg photograph where an acute β^* angle is enclosed by either $+a^*$ and $+c^*$, or $-a^*$ and $-c^*$, i.e. see Figure 2.21. Now the $h01$ festoon of spots can occur on both the zero layer b-axis Weissenberg photograph, or the first layer c-axis photograph. However, $h01$ and $h0\bar{1}$ reflections are not equivalent. If therefore a comparison of $h0l$ reflections on both photographs shows a discrepancy, it is likely that the c-axis photograph was taken about the $-c$-axis. Re-labelling this axis will correct the discrepancy, but may result in the angle β being acute instead of obtuse. A complete structure analysis can be carried out with the β angle acute. (At a later date the atomic co-ordinates may be related to the axes of a conventionally chosen cell having an obtuse β angle at its origin if so desired. Alternatively, but less simply, the complete set of intensity data may be re-indexed after collection, and the structure analysis carried out using the conventional unit cell.)

In the triclinic system, it is necessary to take three zero layer Weissenberg photographs and three first layer photographs, to

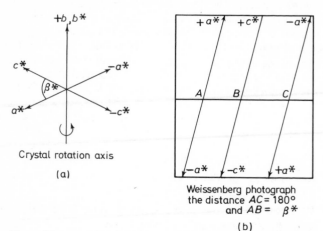

Figure 2.21. The relationship of crystal axes to Weissenberg photographs in the monoclinic case, where the crystal was rotated about its unique +b-axis

confirm that a consistent choice of axes and hence a consistent allocation of indices to reflections has been made. The axial spots on the first layers will correspond to non-axial spots on the other zero layers, i.e.

Axial reflections	*Appear on*
$1k0$	— zero layer c-axis and 1st layer a-axis
$10l$	— zero layer b-axis and 1st layer a-axis
$0k1$	— zero layer a-axis and 1st layer c-axis

A comparison of the relative intensities of these spots on each film will confirm whether or not the axes have been correctly assigned. For example, if the 10*l* spots, i.e. the c^* axis on the first layer photograph of the *a*-axis, do not correspond with the 10*l* festoon on the zero layer *b* photograph, then probably a^* or c^* will need changing to $-a^*$ or $-c^*$. Figure 2.22 shows that if this is done, then the angle enclosed by a^* and c^* will change from acute to obtuse, or vice versa.

Even with a triclinic crystal where an extinction direction does not necessarily correspond to a crystallographic axis, there should

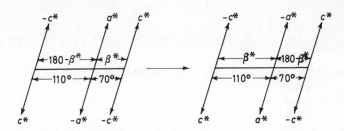

Figure 2.22. The labelling of axes on a Weissenberg photograph to produce either an acute or obtuse β angle

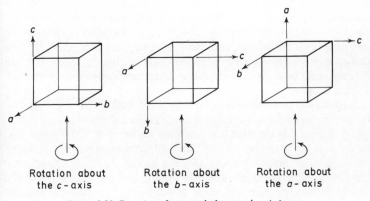

Rotation about
the *c*-axis

Rotation about
the *b*-axis

Rotation about
the *a*-axis

Figure 2.23. Rotation of a crystal about each axis in turn

be little difficulty in finding three consistent crystallographic axes about which to rotate the crystal. Once one axis has been found, two others can usually be found not too far from 90° to the original one. Therefore, the procedure is to find one axis, assume the others are mutually at right angles and rotate the crystal through 90°, bearing in mind its morphology, and also the appearance of the zero layer photograph already taken. This usually results in the crystal being slightly mis-set about another axis, and an oscillation photograph will give the corrections needed to set the crystal accurately about another crystallographic axis.

It is a useful approach to consider all the axes in the monoclinic and triclinic crystal systems as being 90° to one another. A rotation of the crystal through 90° will bring each axis in turn to the axis of rotation, or else, usually sufficiently close to the rotation axis that a correction may be derived from an oscillation photograph to make the crystallographic and rotation axes coincident (see Figure 2.23).

SYSTEMATIC ABSENCES AND SPACE GROUP DETERMINATION

Systematic absences of spots on Weissenberg photographs arise from two main sources; there are those due to the presence of a centred lattice and those due to the presence of screw axes and glide planes in the space group.

It is possible to obtain from the Institute of Physics and the Physical Society[24] a transparent template which can be super-imposed on a zero layer Weissenberg photograph, and the Cartesian co-ordinates of the spots on the photograph can be read off the template. The axes on the template are 90° apart; if the axes on the photograph are separated by a different angle, then each axis can be considered separately, and the template is superimposed first on one axis, when the indices of the festoons cutting the other axis can be read, and then on the other axis when the procedure is repeated. Figure 2.24 shows a drawing of a template.

Systematic absences can be observed either by plotting the co-ordinates of the spots from the film on to a sheet of graph paper, or by drawing in the axes and festoons on the film with Indian ink, or by making a tracing of the film and drawing axes and festoons on it. In most cases, absences are seen merely by observation when the film is superimposed on the template, but the advantage of the above methods is that the same procedure can be carried out for the first layer Weissenberg and a comparison made more easily of zero and first layer photographs, when missing axes and festoons will show up more clearly.

Table 2.2 shows the absences that occur as a result of centred lattices, and Table 2.3 shows the absences that arise from glide-planes and screw-axes[25]. All body-centred crystals have reflections absent where the $h+k+l$ indices are odd, and face-centred crystals give reflections where the indices are either all odd or all even.

Volume I of *International Tables* gives the reflections that are observed with each space group. If space group No. 20, $C222_1$, is

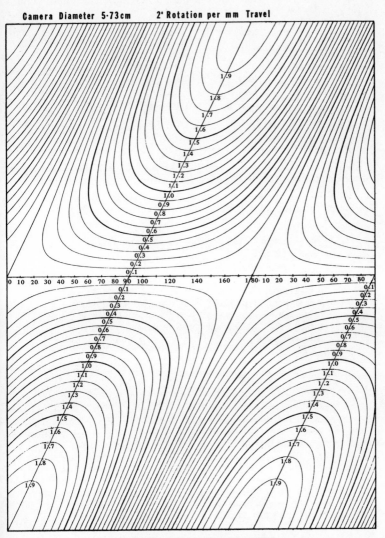

Camera Diameter 5·73cm   2' Rotation per mm Travel

THE INSTITUTE OF PHYSICS & THE PHYSICAL SOCIETY

Figure 2.24. A Weissenberg chart

SOME PRACTICAL CONSIDERATIONS

taken as an example, only the following reflections would be observed in the case where there are eight asymmetric units per unit cell:

hkl	$h+k = 2n$
$0kl$	$(k = 2n)$
$h0l$	$(h = 2n)$
$hk0$	$(h+k = 2n)$
$h00$	$(h = 2n)$
$0k0$	$(k = 2n)$
$00l$	$l = 2n$

i.e. hkl reflections only where $h+k$ is even, are observed. The parentheses show that the condition enclosed is imposed by some other condition. Only $00l$ reflections where l is even are observed.

Volume I of *International Tables* p. 52–55 contains a discussion on the determination of diffraction symbols from possible reflections,

Table 2.2 ABSENCES DUE TO CENTRED LATTICES

Type of lattice	Absences
Body centred	hkl with $h+k+l$ odd
A-face centred (100 face)	hkl with $k+l$ odd
B-face centred (010 face)	hkl with $h+l$ odd
C-face centred (001 face)	hkl with $h+k$ odd
All face centred	hkl with $h+k$, or $k+l$, or $l+h$ odd

Table 2.3 ABSENCES DUE TO GLIDE PLANES AND SCREW AXES

Screw axes and glide planes	Absences
a glide plane, perp. to c, translation $a/2$	$hk0$ with h odd
n glide plane, perp. to c, translation $(a+b)/2$	$hk0$ with $h+k$ odd
d glide plane, perp. to c, translation $(a+b)/4$	$hk0$ with $h+k$ not a multiple of 4
c glide plane, perp. to 110, translation $c/2$	hhl with h odd
Twofold screw axis parallel to c, (2_1)	$00l$ with l odd
Threefold screw axis parallel to c, $(3_1, 3_2)$	$00l$ with l not a multiple of 3
Fourfold screw axis parallel to c, $(4_1, 4_3)$	$00l$ with l not a multiple of 4
(4_2)	$00l$ with l odd
Sixfold screw axis parallel to c, $(6_1, 6_5)$	$00l$ not a multiple of 6
$(6_2, 6_4)$	$00l$ not a multiple of 3
(6_3)	$00l$ with l odd

and gives a comprehensive list of the conditions for reflections when certain symmetry elements are present in the space group. It is not possible in every case to determine the space group of a crystal uniquely from the systematic absences found from photographic data. The reason for this is that only symmetry elements involving translations give rise to systematic absences, and that the diffraction effects produced by a particular point group always contain a centre of symmetry.

Friedel's law states that an X-ray reflection from a set of planes (*hkl*) is the same as that from a set ($\bar{h}\bar{k}\bar{l}$). Even if a crystal is non-centrosymmetric, the diffraction pattern it produces will be centrosymmetric. A Laue group is a group of point groups which become identical when a centre of symmetry is added to those of the group which do not have one. There are eleven centrosymmetric point groups, and hence eleven Laue groups (see Table 2.4).

Table 2.4 THE LAUE GROUPS

Triclinic	—	$1, \bar{1}$	
Monoclinic	—	$2, m, 2/m$	
Orthorhombic	—	$222, mm2, mmm$	
Tetragonal	—	$4, \bar{4}, 4/m$:	$422, 4mm, \bar{4}2m, 4/mmm$
Trigonal	—	$3, \bar{3}$:	$32, 3m, \bar{3}m$
Hexagonal	—	$6, \bar{6}, 6/m$:	$622, 6mm, \bar{6}m2, 6/mmm$
Cubic	—	$23, m3$:	$432, \bar{4}3m, m3m$

It can be seen that there are eleven Laue groups; enclosed by boxes in the above table.

The procedure that is usually followed in determining a crystal space group from diffraction data is first to find the crystal system from Weissenberg photographs. Often the crystal system has already been found from an optical and morphological examination. Systematic absences then give information about the lattice type and symmetry elements that are present, and the symmetry of the X-ray diffraction pattern leads to a knowledge of the Laue group of the crystal.

International Tables for X-ray Crystallography[26] contain a set of tables giving the diffraction symbols of the 230 space groups. (A diffraction symbol consists of first the Laue group, second the lattice type, and third any translation operators from systematic absences). Use is made of these tables in the following way: as much information as possible is obtained about the crystal by both X-ray and non-X-ray methods:

1. The crystal system is determined.
2. Lattice type and symmetry elements are found from systematic absences, and the diffraction symbol is determined.
3. From the tables the possible space groups corresponding to the diffraction symbol can be obtained.

Where only the crystal system is known there are 97 possible different extinction symbols, 38 of which give the unique determination of a single space group. If both the Laue group and the system are known there are 120 possible extinction symbols, 50 of which give the unique determination of a single space group. If the point group of the specimen has been determined by other methods then 186 space groups may be uniquely determined. Of the others, 22 groups are in enantiomorphous pairs, 4 groups are in special pairs and 18 groups are in pairs where the orientation of the point group relative to the lattice must be known in order to distinguish the pairs. In practice 58 space groups are uniquely determinable by X-ray methods.

Information about point groups of crystals can often be obtained from physical properties such as morphology, etch pits, optical examination, optical activity, pyro-electricity and piezo-electricity.

All the available information on a crystal's symmetry will sometimes still not lead to a unique determination of the space group. In these cases the space group is arrived at from a successful solution of the crystal structure. That is to say, intelligent guesswork and trial-and-error lead to a successful structure determination and incidentally give the crystal space group.

THE NUMBER OF MOLECULES, n, IN A UNIT CELL

The value of n is related to the space group of the crystal and is dependent upon the point group and any centring that may be present. Generally the number of molecules is equivalent to the number of asymmetric units in the unit cell, but this is not always the case as an asymmetric unit may contain two or more molecules. The converse is true when the asymmetric unit is less than a whole molecule; the whole molecule being produced by acting on the asymmetric unit with the relevant symmetry operations.

If we take a primitive unit cell as an example and consider the possible point groups in turn, the following values for the number of asymmetric units are obtained:

Point group			Number of asymmetric units
	1		1
$\bar{1}$	2	m	2
$\dfrac{2}{m}$	222	$mm2$	4
	mmm		8

Part 2
Collection and Measurement
of Intensity Data

3 The Weissenberg and the Precession Cameras

THE WEISSENBERG CAMERA

Once the unit cell parameters and the space group (or possible space groups) of a crystal have been determined, the large mass of diffraction data can be collected. For a full three-dimensional solution of a crystal structure, data are collected with a crystal mounted about each of the three crystallographic axes. Three separate crystals can be used, one for each axis, although all the data can be collected from one crystal if necessary. From a crystal mounted about its a-axis, the following diffraction data can be collected (only the one quadrant where k and l are positive is considered here):

$$0kl - \text{zero layer Weissenberg film}$$
$$\pm 1kl - \text{first layer Weissenberg film}$$
$$\pm 2kl - \text{second layer Weissenberg film} \qquad \text{etc.}$$

Similarly, a crystal mounted about the b-axis would have: (h and l positive)

$$h0l - \text{zero layer film}$$
$$h \pm 1l - \text{first layer film}$$
$$h \pm 2l - \text{second layer film} \qquad \text{etc.}$$

and from the c-axis (h and k positive):

$$hk0 - \text{zero layer film}$$
$$hk \pm 1 - \text{first layer film}$$
$$hk \pm 2 - \text{second layer film} \qquad \text{etc.}$$

The sign of the indices above shown as $\pm$ depends upon the sign of the axis of rotation.

Figure 3.1 shows these layers of diffraction data as slices of the reciprocal lattice, and it is obvious that some points will be common to more than one slice, i.e. a reflection from some set of planes will appear on more than one Weissenberg photograph, e.g. the 123 reflection appears on the first layer of data taken about the a-axis, the second layer of data taken about the b-axis, and the third layer of data taken about the c-axis.

When a full set of three-dimensional data has been collected, duplicated or triplicated reflections can be used to put all of the layers of the different axes on a common scale, after the data have been corrected for various factors. These spots are usually averaged and any duplications removed.

Although the axis of rotation can be designated either + or −, the choice must be consistent with a choice of a right-handed set of axes. For this reason, particularly if the crystal belongs to a monoclinic or triclinic space group, it is helpful to be able to relate the crystal and camera geometry to the reciprocal lattice and the reflecting sphere.

Figure 3.2 shows a view looking down the axis of rotation of the crystal, i.e. along the camera spindle axis. The reflecting sphere is

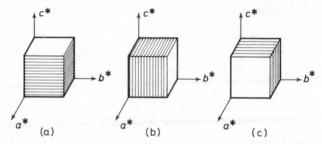

Figure 3.1. Layers of reciprocal space. (a) hkn data where n is the number of the layer, and in these examples is positive. (b) hnl data. (c) nkl data

shown as a circle enclosed by the film. The X-ray beam enters from the left and the reciprocal lattice rotates about an axis parallel to the crystal rotation axis, with its origin at the point where the primary beam meets the sphere of reflection after passing through the crystal. Both the crystal and the reciprocal lattice rotate in an anti-clockwise direction in this example, and when a reciprocal lattice point cuts the sphere of reflection a reflected beam will register on the film. Figures 3.2(a), (b) and (c), show how axial reflections register on the film as the reciprocal axis 1 moves through the sphere of reflection.

It follows that if the direction of translation of the film holder is known as the reciprocal lattice rotates, it should be possible to allocate a right-handed set of axes to the crystal and the reciprocal lattice, and consistently index the Weissenberg photograph that is produced.

If in the example shown in Figure 3.2 the film-holder moves towards the reader, and the crystal is monoclinic and rotates about

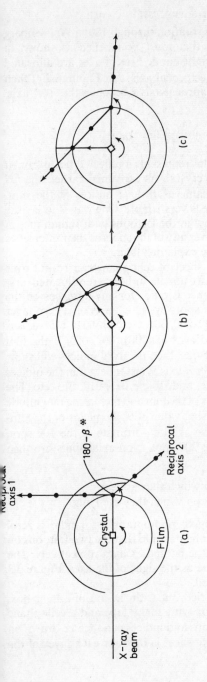

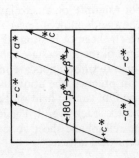

Figure 3.2. Choosing a right-handed set of axes for a monoclinic crystal

its unique $+b$ axis, then after a rotation through 180° a Weissenberg photograph will be obtained that must be labelled as shown in Figure 3.2(d) in order that a right-handed set of axes are allocated where β^* is an acute angle. Reciprocal axis 1 in Figure 3.2(a) then becomes the $-c^*$-axis, and reciprocal axis 2 becomes the $+a^*$-axis.

Allocation of Indices to Weissenberg Film Data

Each spot on a Weissenberg film represents a reflected beam from a particular set of planes in the crystal. As explained earlier, the hkl indices of the plane are the same as the hkl indices of the spot. Indexing all the spots on a film is very simple. One way is to lay the film over the template and draw in the festoons with Indian ink on the film, being careful not to cover any of the spots as their intensities have to be measured, as will be explained later.

Alternatively, a transparent sheet of cellophane can be put over the film and the festoons drawn on it. Indices are allocated first to the axial reflections. Allowance is made for any absences on the film and then, for example, the spots up the a^*-axis on a zero layer Weissenberg are indexed as 100, 200, 300, etc. Similarly, a b^*-axis is labelled 010, 020, 030, etc. and a c^*-axis 001, 002, 003. If the film is not a zero layer Weissenberg one, then the axial indices will show this. For example, on a first layer a-axis Weissenberg film the indices up b^* will be 110, 120, 130, etc. and up c^*: 011, 012, 013, etc. The indexing starts from the equator of the film and increasing magnitude of index corresponds to increasing value of θ. By means of the template it can be checked that the axial co-ordinate of the 100 spot, for example, is half that of the 200 spot. Non-axial spots are then allocated indices:

1. According to the layer on which they appear.
2. According to the festoons on which they are lying.

For example, on a zero layer a-axis photograph the h index is zero. The k and l indices derive from the position that the two festoons on which the spot is situated cut the b^* and c^* axes respectively. The signs of the indices are the same as the signs of the axes. Figure 3.3 illustrates how spots are indexed.

In practice, the above is carried out with the aid of a light box. The film can be held on the box with sticky tape and a cellophane sheet is stuck over it. The festoons and indices can then be drawn on the cellophane sheet. A separate sheet is used for each layer of the reciprocal lattice.

When a set of three-dimensional data are being scaled, the magnitude of the intensity of reflections that appear on more than one photographic film can be used as a check that the indices have been allocated correctly.

When the reflections on higher layer films are being indexed, it is

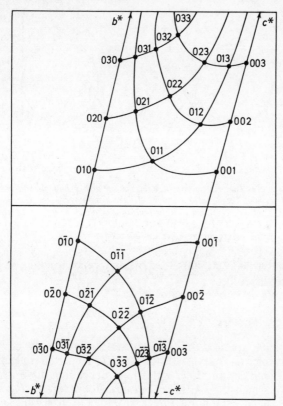

Figure 3.3. The allocation of indices to diffraction spots on a Weissenberg photograph. The diagram represents a zero-layer photograph taken with the crystal rotating about the a-axis. Only the 0kl and 0kl̄ quadrants are shown. On the photograph only the spots would be shown, and the festoons obtained by superimposing a template on the photograph

advisable to bear in mind the indices of the layer immediately below the one being considered, particularly in the case of crystal systems where the axis of rotation does not correspond to the axis of the reciprocal lattice, i.e. the a- and c-axes of monoclinic crystals having the b-axis unique, and all the axes of triclinic crystals. Figure 3.4

shows the relationship between the rotation axis and the reciprocal axis that corresponds to it in these cases.

It can be seen that a distorted version of the zero layer is obtained on higher layer Weissenberg photographs where the rotation axis reflections will also appear on the film, e.g. if the *a*-axis is the rotation

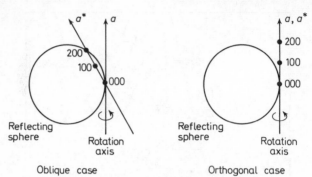

Figure 3.4. The relationship between oblique and orthogonal reciprocal lattices and the reflecting sphere. In the oblique case the reciprocal axis is not co-directional with the real rotation axis, and axial reflections may be obtained on higher layer Weissenberg films when the reciprocal axis lattice points cut the sphere of reflection

axis, then the 200, 300, 400, etc. spots may appear on the corresponding higher layer films.

Multiplicity Factors

The data that are used to solve a crystal structure must be a unique set. Allowance has to be made for the fact that certain reflections are identical with certain other reflections, depending upon the crystal system to which the specimen belongs. Vol. I *International Tables* p. 32–33, gives a table of these factors for each crystal system. If we take as an example the triclinic system, then in general $hkl = \bar{h}\bar{k}\bar{l}$, i.e. the following relationships will be true:

$$h00 = \bar{h}00$$
$$0k0 = 0\bar{k}0$$
$$00l = 00\bar{l}$$
$$0kl = 0\bar{k}\bar{l}$$
$$0\bar{k}l = 0k\bar{l}$$
$$h0l = \bar{h}0\bar{l}$$
$$\bar{h}0l = h0\bar{l}$$
$$hk0 = \bar{h}\bar{k}0$$

$$h\bar{k}0 = \bar{h}k0$$
$$hkl = \bar{h}\bar{k}\bar{l}$$
$$hk\bar{l} = \bar{h}\bar{k}l$$
$$h\bar{k}l = \bar{h}k\bar{l}$$
$$\bar{h}kl = h\bar{k}\bar{l}$$

In drawing up a unique set of diffraction data for a triclinic crystal, it is only necessary to have one of the two equivalent sets of data. In more symmetric crystal systems, the multiplicity is greater. Usually all the available data are measured, and the multiple reflections are then averaged.

When the photographic data are being collected, it is necessary to collect all the unique data available on each layer. This is easily done in the crystal systems of higher symmetry than triclinic. In the triclinic case, it is sometimes necessary to take two Weissenberg photographs of each layer, one with the crystal at a certain setting, and the other with the crystal rotated on its spindle through 180°. Only in this way can all the data be collected. When this procedure is adopted, each of the two sets of data have to be put on a common scale, and this can be done by comparing intensities of reflections that appear on both films as a result of overlap, i.e. the crystal rotates through about 200° and the two sets of data are taken 180° apart.

Collection of Intensity Data by a Weissenberg Camera

The aim of taking Weissenberg photographs is to be able to obtain the magnitudes of the intensities of the diffracted rays. The darkening of the photographic emulsion is proportional to the product of the intensity and the exposure time (see Buerger, M. J., *Crystal Structure Analysis*, p. 78, Wiley), so by measuring the darkness of each spot with a densitometer or a calibrated scale of some kind, a value is obtained for the intensity $I(hkl)$ of each particular diffracted ray. Obviously some spots on the film will be too strong to be measured and some may be too weak to appear. To make sure that all the possible data are collected, particularly higher reflections which may be very weak, a multi-film technique is employed.

Two sets of data are collected for each crystal setting, one a long exposure and one a short exposure. In each case three or four films are mounted together in the film holder and by this means very strong reflections, which would be too strong to measure on one film, will be attenuated to an intensity that can be measured as the ray passes through successive films.

The time taken for the long exposure should be such that the weakest reflections on that particular layer will register on the film. The time for the short exposure can be such that the film containing the strongest reflections (i.e. no attenuation by other films) is identical with the weakest film of the long exposure. If, however, this means that the film containing the most attenuated data of the short exposure still has some spots that are too strong to be measured, then the time of exposure should be reduced.

It is possible to calculate the number of layers of data that can be collected; as they must cut the sphere of reflection, the number will be a function of the wave-length of the radiation and the interplanar spacing of the reciprocal lattice planes perpendicular to the axis of rotation.

The total number of reflections that it is possible to collect can be found by dividing the volume of the unit cell into the volume of the limiting sphere (its radius = $2 \times$ the radius of the reflecting sphere). To obtain the number of unique reflections, allowance has to be made for the multiplicity of the particular crystal system being studied.

Measurement of Spot Intensities using a Calibrated Strip

The opacity, O, of a photographic film is defined by L_0/L, where L_0 is the intensity of the incident beam and L is the intensity of the transmitted beam[27]. A plot of $\log_{10} L_0/L$, which is proportional to the density of silver in the film produced by the incident radiation against E, the exposure time, should give a straight line. This is true up to a limiting density, where quanta of radiation can be absorbed by film grains that have already been sensitised by other quanta of radiation.

Very dark spots on a film may therefore not have a linear relationship between density and exposure time, but as the eye has difficulty in differentiating between varying degrees of blackness of very dark spots, measurements are not actually made where the linear relationship does not apply.

Ideally, the measured intensity should be the integrated intensity of a particular spot, but in practice the maximum intensity of a particular spot is often used, although the relationship between the maximum intensity and the integrated intensity is not linear. Two reflections of the same integrated intensity, one at a low θ value and the other at a high θ value, will have a low and high maximum intensity respectively if the incident X-ray beam is not parallel, due to more focussing of the reflection at the high θ value.

Some Weissenberg cameras have an integrating attachment which gives the film holder a small translation and rotation at the end of its traverse. This produces spots which have an integrated plateau which has been traversed by all the points of the un-integrated spot. However, there is a corresponding increase in the time needed for an exposure.

There are several commercial instruments available which make use of photocells to measure either integrated or non-integrated intensities, and M. J. Buerger in *Crystal Structure Analysis*, Chapter 6, discusses the measurement of intensities by various methods.

A commonly used method of measuring the intensities of diffracted spots on a Weissenberg film is to prepare a calibrated strip on which are spots of known relative intensity, and to obtain a match between one of these spots and the spot being measured by visual comparison. Such a strip can be prepared in the following way.

A spot characteristic of the crystal under examination, of medium intensity, is chosen by examination of a zero layer Weissenberg photograph. The Weissenberg apparatus is arranged so that this spot is in the reflecting position as follows.

The translation of the film holder is geared to the rotation of the crystal so that $2°$ rotation = 1 mm of traverse. Knowing the position of the crystal when the film holder is at the centre of its traverse, it is possible to calculate from measurements on the film what the crystal position must be to produce any particular spot on the film. The crystal is set to oscillate through about $2°$ either side of the angular position of the chosen spot, and the film holder is positioned so that the extreme edge of the film corresponds with the slit of the screens. An exposure is made so that the spot produced by the diffracted beam is just visible when the film is developed. Such an exposure may take a matter of seconds. Using this exposure time to produce the first spot on the calibrated scale, a series of exposures are taken for increasing lengths of time, and after each exposure the camera is translated a few millimetres so that on developing the film a series of spots are obtained a few millimetres apart, having a graded increase in intensity.

The calibrated strip is used in the region where the relationship between darkness of spot and exposure time is linear. A suitable series of spots can be made by counting the number of times the film holder traverses to produce the weakest spot, and using multiples of this to obtain the complete scale. If the weakest spot is allocated an intensity of unity then a suitable range would go from 1 to 30 in steps of integral numbers. M. J. Buerger in *Crystal Structure Analysis* suggests using a geometric progression where the first exposure is for time a, the second for ar, the third for ar^2, etc. Where r is the time needed to produce a spot that is discernibly different from one produced by an exposure for time a.

The advantage of using a reflection from the crystal to make a calibrated scale is that the spots on the scale and the spots being measured have the same shape, which makes for ease of comparison.

Any discrepancies in the calibrated strip will be noticed when similar spots are measured on different films; if the multi-film method is used, a constant scale factor (e.g. $\sim 2 \cdot 9$ for Ilford Industrial G film) will relate similar spots on adjacent films, and a check can be kept that this scale factor is constant.

Scaling of Film Data

When a structure has been solved and the atomic parameters are being refined (this is an iterative procedure which gives the best fit of calculated and observed data by a least-squares process, and is described later), the data are put on an absolute scale by means of a scale factor which is included in the refinement process. However, before this point is reached, all the data must be scaled so that the intensities, $I(hkl)$, are on a common scale. The following scaling procedure would be followed in the triclinic case.

1. The two sets of data which are collected for each layer (except the zero layer) are put on a common scale by comparing spot intensities, $I(hkl)$, where the photographs overlap.
2. If two exposures, one long and one short, are taken for each crystal setting, and in each case a four film pack is used, then seven scale factors are needed to put all the data on one scale.
3. If, for example, five layers of data are collected up each axis, then all fifteen sets of data must be put on a common scale. This is done most simply by relating the intensities of the axial reflections on higher layer films to those on corresponding zero layer festoons.

All the corrections to the data, e.g. Lorentz, polarisation, absorption, etc., should be applied before scaling is carried out.

The first, second, third, fourth and fifth layers of data for the a-axis will show the following axial reflections:

Layer	$b*$ axis	$c*$ axis
1	$I(1k0)$	$I(10l)$
2	$I(2k0)$	$I(20l)$
3	$I(3k0)$	$I(30l)$
4	$I(4k0)$	$I(40l)$
5	$I(5k0)$	$I(50l)$

The $b*$ axes will also appear on the zero layer c-axis photograph, and the $c*$ axes will appear on the zero layer b-axis photograph. Comparison of the spot intensities will give the scaling factors.

A similar procedure for higher layer axial reflections on the films taken about the other two axes (b and c) will result in all the data being scaled.

It should be borne in mind that a unique set of data is required so common reflections should be averaged and surplus data removed during the scaling procedure.

A more comprehensive scaling may be carried out using all reflections that appear on more than one layer. Such a large task is greatly facilitated by the use of a computer, and in this case a limit is usually set so that large discrepancies between similar reflections on different layers, which may be due to mis-measurement or incorrect indexing, are not allowed to interfere with the determination of the scale factors.

All the data have now been assigned an index, the intensity of each reflection has been measured, duplicated reflections have been removed, and all the data are on a common scale. The point has been reached when the crystal structure analysis can be started. Before the various methods of attack are described and the types of computing programs are examined, let us look in slightly less detail at alternative methods of data collection.

Collection of Low Temperature Weissenberg Film Data

Low temperatures are employed during the collection of intensity data for two main purposes:

1. In order to study the crystal structures of compounds which are gaseous or liquid at room temperature.
2. To reduce the amount of thermal vibration displayed by the atoms in a structure and so improve both the quality and quantity of the data collected.

Robertson[28] describes a cryostat for use with liquid hydrogen and discusses the use of both liquid nitrogen and liquid hydrogen as coolants. It is proposed here to describe a modification of this apparatus that was used for the collection of intensity data for *cis*- and *trans*-dichlorodiammine Pt(II) at 120K[12].

When working at low temperatures it is usual to enclose the crystal in a Lindemann capillary tube to protect it in the event that

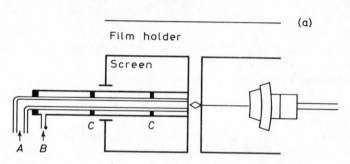

(a)

Film holder

Screen

A-Nitrogen flowing along a tube with evacuated and silvered walls-the cold stream

B-Nitrogen flowing along a glass tube - the warm stream

C-Spacers holding the tubes apart

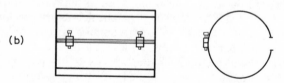

(b)

Figure 3.5(a). The position of the cooling tubes for low temperature Weissenberg photographs. (b) The modified film holder is split so that the cooling tubes need not be removed each time a new film is taken

frost should form on it. Sometimes, as well, liquid nitrogen tends to be pumped in spurts from the cooling tubes if the Dewar flask is overfull, which can be disastrous if the crystal is unprotected.

The Weissenberg camera that was used was operated horizontally, and the only modification to it was the provision of a split film holder so that the cooling tubes could be left permanently in place. This meant, of course, that each half of the film holder was loaded separately with film. Figure 3.5(a) shows the arrangement of the cooling tubes in the camera, and Figure 3.5(b) shows the film holder.

Figure 3.6 shows the liquid nitrogen Dewar flask, of which two are used. One was used as a source of 'warm nitrogen' and no particular insulating precautions were taken with it. The other was

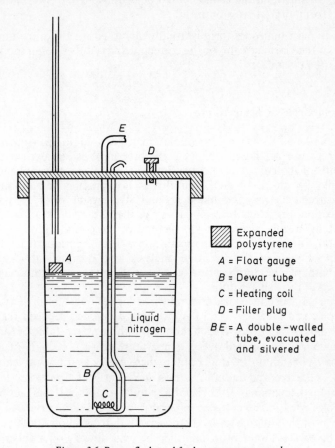

Expanded polystyrene

A = Float gauge
B = Dewar tube
C = Heating coil
D = Filler plug
BE = A double-walled tube, evacuated and silvered

Figure 3.6. Dewar flask used for low temperature work

the source of the 'cold stream', and all the delivery tubes were evacuated and silvered. At the point where the delivery tubes enter the Weissenberg apparatus, the screen and the tubes are concentric, the cold stream tube being inside the warm one.

The liquid nitrogen is removed from the Dewar flasks by means of a small heating coil situated under the exit tube, and the rate of flow of the nitrogen is controlled by the amount of current flowing in this coil.

The purpose of the warm stream is twofold:

1. To flush out any moist air in the camera prior to cooling.
2. To surround the crystal with a dry atmosphere to prevent the formation of frost on it.

Any frost that does form is usually due to either a too low flow-rate of the warm stream, or the cooling delivery tubes being too far from the crystal.

Measurement of Temperature

Copper–constantan thermocouples may be used to measure the temperature of the crystal. One thermocouple is situated in the mouth of the cold stream delivery tube. Ideally this thermocouple should be on the crystal, but this is not practicable, although measurements may be made around the crystal prior to data collection, and a calibration curve can then be made to allow for the temperature gradient from the end of the cooling tube to the crystal. In fact the exact temperature is not important from the point of view of the structure analysis, and a sufficiently good estimate of the temperature is obtained from a thermocouple placed in the delivery tube.

The e.m.f. produced in a circuit by two copper-constantan thermocouples is a measure of the temperature difference between them. One microvolt corresponds to about 1/40th degree. A table of e.m.f. in microvolts generated over a temperature range $+400°-200°C$ by copper-constantan thermocouples where one is at $0°C$, is given on p. 49 of Kaye and Laby's *Physical and Chemical Constants*, 13th Edition (Longmans). By observing similar points for the thermocouples used in the cryostat apparatus, a calibration curve can be drawn of the differences in the observed and the quoted values.

The crystal temperature can then be read directly off a galvanometer, and the cooling of the crystal can be followed quite closely.

Procedure for Cooling the Crystal

The warm stream of nitrogen is allowed to flow at the rate arrived at by trial and error, that will prevent frost forming on the crystal. The rate of flow of the cold nitrogen stream is gradually increased over about one hour so that the crystal is gradually cooled to the required temperature. Rapid cooling may crack the crystal, especially if a rigid adhesive has been used to attach it to the tube. Rapid warming may also crack the crystal, and should be guarded against.

Once the crystal is at the required temperature, it is advisable to check that it is still correctly oriented by taking a setting photograph before proceeding with Weissenberg photographs.

The technique that is used with gaseous or liquid samples is to freeze them completely in a capillary tube and then allow them to warm up slowly until only a small solid portion is left. This may then be used as a seed from which to grow a single crystal by further cooling.

THE PRECESSION CAMERA[29]

The precession camera is a flat plate moving-crystal moving-film camera which produces undistorted pictures of a chosen layer of a reciprocal lattice. It was designed by Buerger, and a monograph by him[30] discusses the theory and practice of its use. The area of reciprocal space which is reproduced on the film is much more limited than that produced by a Weissenberg camera, and for this reason the two cameras are usually used in conjunction. The precession camera has an important application in the confirmation of space group symmetry, and the use of polaroid film enables very rapid collection of data.

Camera Operation

Consider a reciprocal lattice net perpendicular to the incident X-ray beam at the origin of the sphere of reflection, i.e. at the point 0 where the primary beam passes out of the sphere (see Figure 3.7(a)). As the crystal rotates about an axis perpendicular to the beam the reciprocal net also rotates about 0 to give an arrangement as shown in Figure 3.7(b). Figure 3.7(c) shows the position of the film which is parallel to the reciprocal lattice net. The cone of reflections produced by the circular intersection of the lattice net with the sphere of reflection is shown projected on to the film, AB. DX is a crystal axis which initially is parallel to the beam. In Figure 3.7(c) it is displaced on angle ϕ. Movement of the crystal so that X rotates about the incident beam from $X \rightarrow X' \rightarrow X$ in Figure 3.7(d) while ϕ is kept constant, will result in the intercepting circle

CD rotating about the origin. If the film is kept always parallel to the reciprocal lattice net then it will precess about the point *B* as the axis *DX* rotates about the beam. As the reciprocal lattice points cut the sphere of reflection an undistorted image of the reciprocal net is produced on the film.

In the example only the zero-level lattice net was considered, but in fact other levels will also be cutting the sphere of reflection and a screening device which also precesses must be inserted, so that

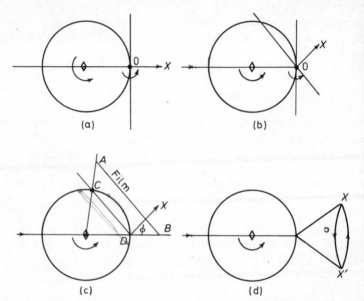

Figure 3.7. The geometry of operation of the precession camera

only one lattice level of data is collected at a time. Figure 3.8 shows a typical precession photograph.

Aligning the Crystal

Although a crystal can be aligned solely using a precession camera, it is simpler to ascertain the position of the crystal axes with respect to the goniometer head using Weissenberg photographs, and then transfer the crystal on the goniometer head to the precession camera. The problem of setting the crystal so that an axis is parallel to the incident X-ray beam varies according to the crystal class of the crystal.

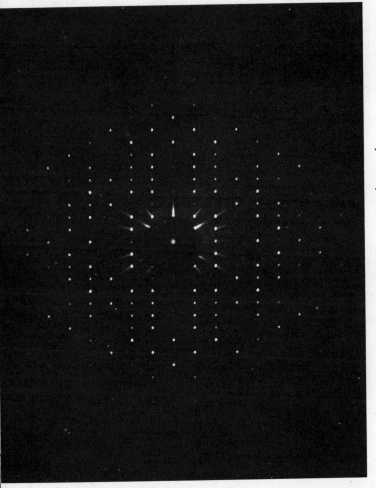

Figure 3.8. Typical precession photograph

In orthogonal systems the real and reciprocal axes coincide and the position of the rotation axis of the crystal on a Weissenberg camera, where a reciprocal axis line cuts the equator of the Weissenberg film will be that position where the reciprocal axis is vertical on a horizontal Weissenberg camera. Rotation of the crystal through 90° will turn the axis parallel to the beam.

In monoclinic crystals mounted on the goniometer head about the unique b-axis, when the a^*-axis is vertical the c-axis is parallel to the beam, and when the c^*-axis is vertical the a-axis is parallel to the beam. These are the usual settings adopted for monoclinic crystals. If it is required to take precession photographs with the c- or a-axis as the precession axis (lying parallel with the X-ray beam) then a^* or c^* are used as the crystal rotation axes respectively; that is, the a^* or c^* axis coincides with the rotation axis of the goniometer head.

With triclinic crystals, a reciprocal axis is arranged perpendicular to the beam and then the degree of mis-setting of the real axis from a position parallel to the beam is obtained photographically, and the necessary corrections made.

When a crystal is being aligned on the camera, unfiltered radiation is used with no screen and it is preferable to have the goniometer head arcs parallel and vertical to the incident beam. A perfectly aligned crystal will produce a circle of spots lying on white radiation streaks which is perfectly symmetrical about the beam stop image, i.e. the position where the incident beam cuts the film. Mis-alignment results in the circle being eccentric about the beam stop image, and measurements of this eccentricity allow corrections to be made so that the crystal may be correctly aligned.

4 Automatic Data Collection and Measurement

A more stable source of X-rays is required when X-ray diffraction data are collected by counter methods rather than by film methods, as the time of collection is much shorter and any fluctuation in the intensity of the X-ray source will seriously affect the quality of the data.

THE WEISSENBERG GEOMETRY DIFFRACTOMETER[31]

Introduction

Figure 4.1 shows a symbolic representation of a Weissenberg geometry diffractometer.

It differs from the camera in that the film holder is replaced by a counter which can be rotated about the crystal rotation axis in order to monitor any of the cones of reflection of the crystal. This rotation is limited by the geometry of the apparatus, and it is not possible to cover a full 360° because of the interference of the X-ray tube shield and the diffractometer base.

Scintillation and proportional counters give pulses whose heights are proportional to the energy of the radiation falling on the counter and inversely proportional to the wavelength of the incident radiation. Selection of pulse heights of a certain magnitude is equivalent to using monochromatic radiation, and the pulses may be counted or used to charge a capacitor so that its charge at any time is proportional to the number of counts registered per second. A potentiometric recorder can then be operated with this charge, so that an X-ray beam reflected by a particular set of planes on falling on the counter will produce a displacement on the potentiometer recorder.

When a measurement is made of a diffracted beam on any particular layer (i.e. on any cone of diffracted rays), there are two angular settings which are used to bring the counter and diffracted beam into coincidence: ϕ, which is the angle of rotation of the crystal about its spindle (ϕ has an arbitrary zero position usually chosen so that a reciprocal axis is parallel to the incident X-ray beam);

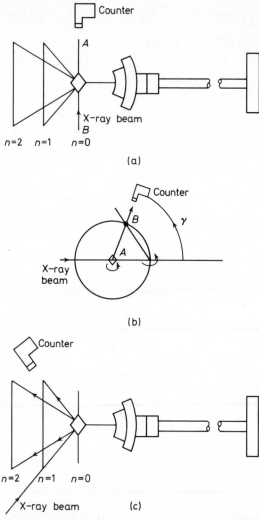

Figure 4.1. A symbolic diagram showing the use of a Weissenberg geometry diffractometer for measuring zero and higher layer cones of diffracted X-rays. (a) The measurement of zero-layer reflections. The counter can rotate around the circle shown projected on AB. (b) The zero layer circle of reflections, seen looking down the crystal rotation axis. AB is a diffracted ray produced when reciprocal lattice point B cuts the sphere of reflection. (c) The measurement of higher level cones of data. The arrangement is similar to that of photographic Weissenberg methods

and γ the angle of rotation of the counter about a cone of reflections measured from the horizontal plane containing the incident X-ray beam and the axis of rotation of the crystal. By comparison with a Weissenberg zero layer photograph γ can be seen to be 2θ (θ = Bragg angle), i.e. a measure of the vertical distance of a spot on the film from the equator. See Figure 4.1(b). ϕ is a measure of the horizontal distance.

The procedure for making an intensity measurement is as follows:

1. The counter is rotated through an angle γ, so that it is in a position to receive a particular diffracted ray when the planes producing it are at a particular angle ϕ.
2. The crystal is set at an angle $\phi - \Delta$, where Δ is a small angle ($\sim 1\cdot5°$) that just removes the crystal from its reflecting position.
3. A background count is made for a known time by the counter, and the crystal is then rotated through 2Δ, i.e. it passes through the reflecting position. The rate of rotation is known so that the time of the count is known.
4. A second background count is known.

All the data collected are both written on a printer output and punched on a paper-tape output. Usually with manual operation the operator punches in the indices of the plane for which data are being collected before the above procedure is followed. One line of output might then appear as:

(Indices)	(Time(s))	(Count)	(Time(s))	(Count)	(Time(s))	(Count)
6 1 6	20	150	40	1068	20	144

To obtain the angular settings for each reflection, the unit cell parameters must be known very accurately ($<0\cdot1\%$); see later. These are obtained from angular measurements made on the diffractometer, consequently there must be no backlash in the apparatus, e.g. in the goniometer-head arcs, which will lead to inaccurate cell parameters.

When the intensity data are collected from layers other than the zero layer, the angles μ and v must be known. In the equi-inclination arrangement which is usually used, and which is the only one discussed here, $\mu = v$. μ is the angle of inclination of the incident X-ray beam to a plane normal to the crystal rotation axis. v is the horizontal angle the counter makes with the plane normal to the crystal rotation axis.

The preliminary work necessary before the data can be collected resolves itself into two sections. One is the accurate alignment of the crystal on the spindle, and the other is the measurement of angular values of ϕ and γ from the accurately aligned crystal so that the unit cell parameters can be obtained and used to find accurate angular settings for every reflection.

Alignment of the Crystal

The crystal, which has already been aligned on the goniometer head rotation axis by photographic means, is transferred from the Weissenberg camera to the diffractometer. A lens attachment which fits over the X-ray collimator allows visual centring of the crystal on the rotation axis of the instrument by means of adjustments to the goniometer head slides.

The procedure to be followed differs for different crystal systems, and falls into two classes. The first contains orthogonal crystal systems, and includes monoclinic crystals mounted about the unique axis. The second is applicable to oblique crystal systems. In both cases it is useful to have a zero-layer Weissenberg photograph taken about the crystal's rotation axis and to know the relationship between the position of the arcs and the appearance of the spots on the film.

Orthogonal Crystal Systems (Crystals rotating about the b-axis)

A $0k0$ reflection with a μ value between 10 and 20° is used to align the crystal. With $\gamma = 0$, and $\mu = -\nu = \sin^{-1} 0.5\,kb^*$, a reflection should register on the counter throughout the rotation of the crystal through 360°. The counter collimator apertures are set at about 2 mm initially, and reduced to a pinhole once the reflections have been found.

One of the goniometer head arcs is set parallel to the X-ray beam and with the X-rays turned on this arc is adjusted to give a maximum reading on the counter (observed by means of the recorder). The arcs are rotated through 180° and any drop in the count rate is due to a mis-setting in μ or in the arcs. μ is adjusted to give a maximum count rate. The true value of μ is then the mean of these two μ readings. With this value of μ the arc is then adjusted to give a maximum count rate, and on rotating to a position 180° away in ϕ no drop in count rate should be observed. The crystal is next rotated through 90° and the other arc adjusted to give a maximum count rate. The above procedure may need to be repeated. Once the crystal is accurately aligned there should be no large change in count-rate during the rotation of the crystal through 360° that cannot be explained by the crystal shape.

Oblique Crystal Systems

Consider a crystal belonging to an oblique system mounted so as to rotate about its *a*-axis. It is easier to apply corrections to the arcs if one of the crystal's reciprocal axes, say b^*, is parallel to one of the arcs. The crystal is aligned using zero layer reflections so that $\mu = -v = 0$, and only γ and ϕ need to be considered. It

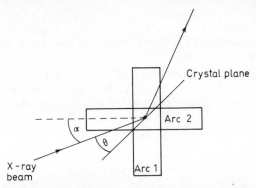

Figure 4.2. Corrections to the arcs used on a Weissenberg geometry diffractometer (plan view)

is convenient to have a zero layer Weissenberg photograph available whose orientation during exposure can be related to ϕ, the position of the crystal on its rotation axis.

A reflection from the $-b^*$ axis with a value of γ between 60 and 90° is chosen, the counter slits are adjusted to about 2 mm, the X-rays are turned on and ϕ is slowly rotated until the reflection registers on the counter. The counter slits are reduced to pinholes, when it will be necessary to adjust v to obtain the reflection in the counter. The magnitude of the mis-setting is given by:

$$\delta = \frac{\Delta v}{2} \cdot \frac{1}{\sin \theta} \quad \text{where } \Delta v = |v_{\max} - v_0|$$

If the plane producing the reflection is parallel to one of the arcs the correction is applied to it, otherwise it reduces to two components, one to be applied to each arc (see Figure 4.2):

$$C_1 = \delta \sin (\theta + \alpha)$$
$$C_2 = \delta \cos (\theta + \alpha)$$

The smaller value is applied to the arc nearer to the plane being used.

The above procedure is repeated for a second reflection approximately 90° away from the first arc. Finally, a check is made that the value of v is the same for a maximum count rate at ϕ values 180° away from the above two reflections, i.e. that $0kl$ and $0\bar{k}\bar{l}$ reflections occur at the same value of v.

Calculation of Cell Dimensions

Orthogonal Crystal Systems

Zero layer reflections lie on a circle at right angles to the axis of rotation of the crystal, with the crystal at the centre of the circle. The relationship between a Weissenberg photograph and ϕ and γ measured on the diffractometer is shown in Figure 4.3.

For the zero layer $\mu = -v = 0$. The angle γ of an axial reflection at large θ value is measured from the photograph, and ϕ is rotated

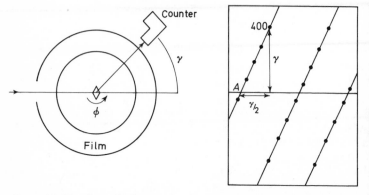

Figure 4.3. The relationship between ϕ and γ of a Weissenberg geometry diffractometer and the zero-layer Weissenberg photograph. The ϕ-value at A on the film's equator is given, for example, by the 400 reflection's γ value, as the ϕ value at A = the ϕ-value at 400–$\gamma/2$

to the correct position for the reflection to fall in the counter which has its collimator set at 2 mm. Once the reflection is found a pinhole collimator can be used to give γ more accurately. It is checked that a 180° rotation of ϕ results in the symmetry related reflection falling in the counter and giving the same peak height. If this does not happen the crystal needs re-aligning. The dimension of the reciprocal axis is easily found from the value of γ where the reflection occurs.

The above procedure is repeated for reflections on the other zero layer reciprocal axis.

The dimension of the third reciprocal axis, which will be parallel to the rotation axis of the crystal, is obtained by measuring μ for a reflection on a higher layer, e.g. if this is the b^* axis, then a $0k0$ reflection where $\gamma = 0$ is used.

$$\mu = \sin^{-1} 0{\cdot}5 \, (kb^*) \quad \text{where } k = \text{the layer number}$$

If the crystal being examined belongs to the monoclinic system and rotates about its b-axis, then Figure 4.3 shows how the β^* angle can be obtained. The values of ϕ_1 and ϕ_2 for two reflections, one on each reciprocal axis lying in the plane perpendicular to the axis of rotation of the crystal, are obtained. $\phi_1 - \gamma/2$ gives the ϕ value where the reciprocal axis cuts the equator of the film, and $\phi_2 - \gamma/2$ gives the ϕ value where the other reciprocal axis cuts the equator. The difference between these two values gives β^*.

Oblique Crystal Systems

When a crystal belonging to an oblique crystal system has been aligned on the diffractometer, the two reciprocal axes in the plane perpendicular to the crystal rotation axis and the angle between them, can be found in the same way as described above for a monoclinic crystal mounted about the b-axis. One way of obtaining accurate cell dimensions would be to repeat this procedure with the crystal mounted about each of its three axes. Alternatively, a least-squares procedure may be used as follows:

1. Cell dimensions obtained by photographic methods are used to generate the γ, ϕ, and μ ($\mu = -v$) angles for every reflection.
2. The zero value of the ϕ scale is found by locating a reflection on the zero layer whose indices are known, and locking the ϕ scale to the calculated value.
3. Other reflections on the zero layer are then looked for at the calculated angles, and confirmation obtained that the instrument is functioning properly.
4. Strong reflections on several layers then have their μ, ϕ and γ value measured accurately (using a pinhole slit in the counter collimator).
5. These values are then used in a least-squares procedure to obtain more accurate cell dimensions from the ones obtained by photographic methods.

One method for doing this is to calculate θ values for each reflection and use the relationship

$$\frac{4\sin^2\theta}{\lambda^2} = h^2a^{*2} + k^2b^{*2} + l^2c^{*2} + 2hka^*b^*\cos\gamma^* +$$

$$2klb^*c^*\cos\alpha^* + 2lhc^*a^*\cos\beta^*$$

in a least-squares fitting procedure.

M. J. Buerger, in *Crystal Structure Analysis*, shows that the setting angles μ, ϕ and γ can be calculated from the following equations:

$$\mu = \sin^{-1}(\zeta/2)$$

$$\phi = \frac{\gamma}{2} + 90° - \psi$$

$$\gamma = 2\sin^{-1}(\xi/2\cos\mu)$$

Where ζ, ξ, and ψ are the cylindrical co-ordinates in reciprocal space of a lattice point $P(hkl)$. A table of relationships of ζ and ψ for different crystal systems is given on p. 124, *Crystal Structure Analysis*.

Arndt and Willis also discuss the equi-inclination case in *Single Crystal Diffractometry*, p. 31 (Cambridge University Press).

Procedure Used to Collect Intensity Data

Once the setting angles for every reflection have been calculated from the unit cell parameters, data collection can begin. The data are collected a layer at a time, starting with the zero layer where $\mu = 0$ and working up through the layers to μ_{max}.

Manual operation can be tedious, and the greater the degree of automation available, the easier the operator's task. A typical manual procedure would involve the setting of $\phi + \Delta$ and γ for each reflection, and punching out the reflection indices on the paper tape output from a printer. The instrument would then read a background count, scan the reflection by rotating ϕ, and read a second background count. This information, together with the relevant times, would then be printed out as both a written record and a paper tape. In this way, all the information is obtained in a suitable format on a paper tape ready for computation.

An improvement to the above procedure is to control the instrument by means of a small computer. Additional stepping-motors must be added to the instrument and the whole of the data collection can be made automatically, ϕ and γ being set by the computer. The angle μ for each layer must still be set manually.

THE FOUR-CIRCLE DIFFRACTOMETER

Figure 4.4(d) shows the geometry of a four-circle diffractometer.

There are two approaches to the rapid collection of data that removes the tedium from the work of the crystallographer. One is

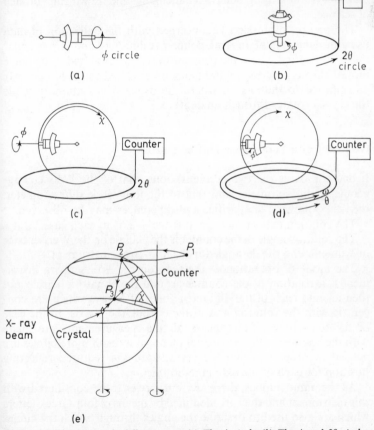

Figure 4.4. The four-circle diffractometer. (a) The ϕ-circle. (b) The ϕ and 2θ circles. (c) The ϕ, χ, and 2θ circles. (d) The ϕ, χ, ω and 2θ circles. (e) Rotation of the crystal through ϕ degrees moves the reciprocal lattice point from P_1 to P_2 on the reflecting sphere. Rotation of the χ circle then moves the point to P_3 when the diffracted ray lies in the equatorial plane (in which the counter moves)

to collect the data photographically and measure them using a computer controlled scanner. The other is to collect the data using a computer controlled four-circle diffractometer[32], i.e. the fully automatic diffractometer. This latter procedure is probably the nearest approach there is to factory-belt production of crystal structures, and assuming that suitable computing facilities are available and enough problems exist to be tackled, there should be little difficulty in solving a medium-sized structure in a month. However, a research program based on such a rapid production of molecular structures would undoubtedly run into some of the problems that make crystallography such a challenging and rewarding branch of science.

The present discussion is concerned with the collection of data using moving-crystal, moving-counter techniques, and in particular the $\omega/2\theta$ method where the detector moves through twice the angle moved through by the crystal (sometimes called the $\theta/2\theta$ method), as opposed to the ω-scan where the detector is stationary while the crystal rotates through an angle ω.

Geometry of a Four-Circle Diffractometer

If one compares the two configurations, that of the Weissenberg-geometry diffractometer and that of the four-circle diffractometer, then the following similarities and differences may be observed.

The rotation axis of the crystal is the ϕ-axis in each case.

The rotation angle of the counter is the γ-angle on the Weissenberg instrument, and the 2θ-angle for the four-circle instrument.

The mode of operation of the scan on the Weissenberg instrument is a rotation of the counter around the equatorial circle and then around each of the diffracted cones of X-rays which are concentric with the rotation axis of the crystal, the ϕ-axis. In the case of the four-circle diffractometer all the measurements are made with the counter in the equatorial plane; as a result more circles are needed in order to bring the crystal into the required reflecting position for each of the *hkl* planes in turn.

As the name implies, there are four circles to be considered with this instrument and they are identified by means of four Greek letters which are also used to describe the angles through which the circles may be moved. To obtain a clear idea of how these circles are used, it is necessary to consider the purpose of the instrument and the way in which X-ray reflections are produced. A crystal lying in a beam of X-rays produces a Bragg reflection when a reciprocal lattice point lies on the sphere of reflection. A diffractometer must therefore be able to so arrange the geometry of the crystal and counter with respect to the beam that every set of crystal planes will reflect into the counter in turn, i.e. the reciprocal lattice point of every set of planes must be brought in turn to a point on the reflecting sphere and the reflected beam must then be directed into the counter.

The first conditions are brought about by means of the χ and ϕ circles. The ϕ circle is in fact the circle of rotation of the crystal at right angles to the spindle axis on which the crystal is mounted, i.e. this is the same as the angle ϕ described earlier when the Weissenberg camera and diffractometer were considered. Figure 4.4(a) shows the ϕ circle, and the crystal may be mounted on a set of arcs or a rigid spike.

When the $\omega/2\theta$ scan is used all the reflection data are measured with the counter rotating in a horizontal plane. This circle of rotation is the 2θ circle. If the crystal spindle axis is at right angles to this circle then the zero layer reflection data of the rotation axis of the crystal (if the crystal is mounted about a crystal axis, which it need not be) will all lie in this circle, and rotation of the counter through different angles 2θ will allow measurements to be made of all the zero layer data, in the same way that zero layer data is collected on a Weissenberg geometry diffractometer by rotating the counter through an angle γ. Figure 4.4(b) shows the 2θ circle at right angles to the crystal spindle axis.

In this arrangement ϕ and 2θ are concentric. It is obvious that using the diffractometer in this way limits the use to which it may be put. Use is made, therefore, of the χ circle, so that any reflection which has been brought into coincidence with the reflecting sphere by means of the ϕ circle may then be brought into the horizontal plane; the reflection may then be measured by setting the counter to the appropriate angle on the 2θ circle. Figure 4.4(c) shows the χ circle in relation to the ϕ and 2θ circles. The ϕ circle, i.e. the crystal spindle, is mounted inside the χ circle. In the $\omega/2\theta$ scan procedure the χ circle is always at right angles to the horizontal 2θ circle.

During the measurement of a reflection the counter moves through an angle 2θ and at the same time the χ circle rotates about a vertical axis through an angle $\omega = \theta$. This is done by means of the fourth circle, the ω circle, which is concentric with the 2θ circle and on which the χ circle is mounted. Figure 4.4(d) shows the arrangement of the ϕ, χ, 2θ and ω circles. The usual ω setting is that where the plane of the χ circle bisects the incident and reflected beams.

To sum up, the ϕ and χ circles bring a set of reflecting planes to that position where the reflected beam lies in a horizontal plane.

The ω and 2θ circles cause the crystal to rotate through a reflecting position and allow the reflected beam to fall on the counter, respectively. Figure 4.4(e) shows how the ϕ and χ circles operate with respect to the sphere of reflection.

Other units which are used in conjunction with the 4-circles are usually:

1. A stable source of X-rays.
2. Scalers, which count the detector pulses.
3. A chart-recorder coupled to the counter which shows the rate of count.
4. Attenuators which may be inserted in front of the counter.
5. A means of supplying the necessary setting angles for each reflection to the instrument.

Data Collection

The continuous collection of reflection data is carried out by supplying the instrument with the calculated setting angles ϕ, χ, ω, 2θ, in a suitable sequence for all the crystal planes. This can be done by direct computer control or by the use of a 'black box', which operates on information read off punched cards. In the latter case a typical procedure would be to have two cards for each reflection. The first one would contain the *hkl* indices of the reflection, and the ϕ, χ, ω, $\Delta\theta_1$, and $\Delta\theta_2$ angular settings. $\Delta\theta_1$ and $\Delta\theta_2$ represent the angles to be scanned by the counter. The angles on the ω circle are directly related to the 2θ angles as $\omega = \theta$. The card reader reads the indices and angles and the necessary pulses are fed to the stepping motors which rotate the circles. The initial background count is made at $2\theta - \Delta$, the peak-count is made over $2\theta \pm \Delta$ and a second background count is made at $2\theta + \Delta$.

The data that have been collected, together with the information on the first card, are then all punched out on the second card. The procedure is repeated until there are no more cards. It is only necessary to sort the cards to obtain a complete set of data with which to start the structure analysis.

Various safety measures can be built in to the system. For example, at specified intervals a set of standard reflections are measured so that any deterioration in the crystal or any instrumental errors may be detected. These standard reflections may also be used to scale sets of data from different crystals etc.

The above description of the procedure using the $\omega/2\theta$ scan method is not detailed, as any procedure depends to a large extent on the equipment and computing facilities that are available, and any system reflects the wishes of the person operating the instrument. The overall procedure is, however, quite general and consists of the following steps:

1. A crystal is chosen which has, if possible, identifiable faces.
2. Zero and upper layer Weissenberg photographs are taken of the crystal, so that strong reflections may be selected to use in setting up the crystal on the diffractometer.
3. The space-group and cell parameters of the crystal are obtained by photographic methods.
4. The crystal is aligned on the instrument and more accurate cell parameters obtained.
5. The setting angles ϕ, χ, ω, 2θ, are obtained from the new cell parameters for the strong reflections selected from the Weissenberg films.
6. The measured values of ϕ, χ, ω, 2θ for these reflections, where the maximum count rate is obtained, are used in a least-squares procedure to calculate very accurate cell parameters.
7. These very accurate cell parameters are used to generate ϕ, χ, ω, 2θ for all the reflections that are to be measured, allowance being made for the systematic space-group absences.

Let us now look in some more detail at the procedure to be followed in setting up the crystal on the diffractometer, and later we will consider the procedure that is followed on the fully-automatic four-circle instruments.

Aligning the Crystal[33]

The crystal is first aligned visually using a microscope and cross-wires that are attached to the instrument, in the same way that a crystal is aligned on a Weissenberg camera.

Further alignment may then be carried out at two positions on the circle. The first, $\chi = 180°$, is that position where the crystal points vertically downwards. The second, $\chi = 90°$, is one of the two possible positions where the crystal rotation axis is horizontal. Reflections are identified using a knowledge of the crystal orientation during the production of a Weissenberg photograph from which the 2θ values are obtained.

$\chi = 180°$

The crystal is considered aligned when two reflections, 90° apart on the ϕ circle, can be made in turn to fall in the centre of the counter aperture by rotating the crystal on its ϕ axis. Initially reflections are chosen having low 2θ values. The counter aperture may be masked by either of two sets of shutters; one set horizontal and one set vertical. In this way the right and left half, or the top and bottom half of the aperture, may be masked. When a reflection falls exactly on the centre of the aperture, operation of each of the shutters in turn will result in the same peak count being observed.

The chosen reflection is allowed to enter the aperture and its peak count value observed on a chart recorder is made a maximum by adjustment of 2θ and ϕ. Use is then made of the aperture shutters, and adjustment of the goniometer head arc which lies closest to $90°$ to the reflected beam should correct mis-setting of the crystal in this orientation. A second reflection is then chosen at $90°$ on the ϕ circle to the first one, and the procedure repeated. If, for example, a 200 reflection and a 020 reflection are used, then the a^* and b^* axes should be parallel with the arcs of the goniometer head. If they are not, then after applying a correction using the second reflection, it will be necessary to repeat the procedure until the optimum conditions are obtained.

The process can then be repeated using reflections having larger 2θ values which will be more sensitive to mis-setting.

$\chi = 90°$

If the crystal rotation axis coincides with a reciprocal lattice axis, then the 2θ angle may be set so that an axial reflection (which is necessarily insensitive to change of ϕ) may be measured in that position where one of the goniometer head arcs is in the equatorial plane. 2θ is adjusted to give a maximum peak count, and the crystal is rotated through $180°$ when the same peak count should be observed. The horizontal arc is adjusted to make the counts equal. The procedure is then repeated with the other arc horizontal.

Checking ω and 2θ

With χ at $90°$, 2θ is adjusted so that a chosen reflection will give a maximum peak count. ω is then rotated to either side of the peak to a point where half the peak count is observed. The average of the two ω readings give the true peak position on ω.

Use is next made of the counter aperture shutters to check the 2θ setting. First the vertical shutters are used to find which half of the aperture is receiving the reflected beam. The counter is moved in 0·1° steps on 2θ so that the aperture centre moves towards the beam. (ω moves $2\theta/2°$ in the opposite direction). By using the shutters in this way the reflected beam is made to fall exactly on the centre of the counter aperture.

At this point, the values of ϕ at 0° and 180°, and 90° and 270°, can be rechecked, using only one half of the counter aperture, i.e. one of the vertical shutters is kept closed. The value of ω is then rechecked, and the whole cycle repeated so that consistent values are obtained for ω and 2θ.

A final check is to measure the peak count at -2θ without moving ω. This should be the same value as that obtained at $+2\theta$.

Determination of Cell Parameters

Once the crystal has been aligned the lattice parameters are obtained from the observed values of 2θ for chosen reflections. Using the $\omega/2\theta$ scan where ω and 2θ are geared in a $1:2$ ratio, the peak position of a reflection hkl on ω is noted, and ω is then rotated so that the reflection is measured from the back of the hkl planes, i.e. the $\bar{h}\bar{k}\bar{l}$ reflection. 2θ is given by the difference of the ω readings. A least-squares procedure using the equation given earlier for the Weissenberg geometry diffractometer will then give the cell parameters from which the setting angles ϕ, χ, ω, 2θ for all the reflections may be obtained.

The Background Count

Arndt and Willis[34] discuss the various factors which contribute to the background count and which may introduce both systematic and random errors in the observed intensity of an X-ray reflection.

In the present account only the counting statistics and the relation between the peak height and the associated background count will be mentioned.

The integrated peak intensity is given by the difference between the peak count, N_p, and the background count N_b.

$$I \propto N_p - N_b$$

If σ_p and σ_b are the standard deviations of the two counts then

$$\sigma_I = (\sigma_p^2 + \sigma_b^2)^{\frac{1}{2}}$$

Arndt and Willis[35] give a table of percentage standard deviations, $100\sigma(N_p - N_b)$, for different values of N_p and N_b. They also derive the following expressions for the counting times spent on peak and background in order to minimise the percentage statistical uncertainty in I.

$$T_1 = (K^{\frac{1}{2}}/1 + K^{\frac{1}{2}}) . T$$

T = overall time, and K is the ratio of the counting rates

$$T_2 = (1/1 + K^{\frac{1}{2}}) . T$$

For weak reflections where $N_p \sim N_b$, $K = 1$ and equal times should be spent on peak and background counts. If the peak to background ratio is large then only a small proportion of the total time should be spent on the background.

It is interesting to note that the total time spent on measuring a weak reflection with peak to background ratio 2:1 should be 34 times longer than the corresponding time for a peak to background ratio of 10:1.

In practice, a background count is made both before and after the peak count, and in the case where the time spent on each background is half the time spent on the peak count, the integrated peak intensity is $N_p - (N_{b1} + N_{b2})$ where N_{b1} and N_{b2} are the background counts. It is emphasised that the background level should be maintained as low as possible during data collection.

Unobserved Reflections

It is common practice to exclude unobserved reflections, as opposed to systematically absent reflections, from the refinement process of a structure, even though theoretically they should be included if correct weights can be allocated to them.

During the process of data collection a reflection is usually considered to be unobserved if its net count is less than 1–2 times the standard deviation of the background count.

Weighting of Data[36]

The correct weight, W, to be assigned to a reflection is the reciprocal of the variance of the reflection, i.e.

$$W = 1/\sigma^2 \qquad \text{where } \sigma \text{ is the standard deviation}$$

For counter data, σ can be estimated from counting statistics and

$$\sigma = \sqrt{N} \qquad \text{where } N = \text{the number of counts}$$

If the two background counts are made for half the time taken for the peak count, then the net peak count, I, is given by:

$$I = N_p - N_b \qquad \text{where } N_b = N_{b1} + N_{b2}$$

The standard deviation of I is then given by:

$$\sigma_I = \sqrt{(N_p + N_b)}$$

This expression allows only for the statistical errors in the counting procedure, and further terms may be introduced to allow for other errors that may arise during the measurement of the intensity data.

SOME EXAMPLES OF FULLY AUTOMATIC FOUR-CIRCLE DIFFRACTOMETERS

Most modern four-circle diffractometers are fully automatic, that is, their operation is fully controlled by a small computer. A crystal positioned at the centre of the instrument can be oriented and its cell parameters determined by means of programs supplied with the instrument. Four examples of such commercial instruments are the Philips computer controlled single crystal X-ray diffractometer; The Siemens-Hoppe automatic single crystal diffractometer; the Stoe automated four-circle diffractometer system STADI 4; and the Hilger and Watts Y290 diffractometer.

The Philips PW 1100

The PW 1100 (see Figure 4.5) consists of a 3 kW generator plus incident beam assembly, the four-circle diffractometer, electronic hardware and the associated controller programs or software. Data processing software is not included in the system.

The instrument is controlled by a fully integrated 8K, 16 bit computer, the Philips P9201, which is supplied with pre-programmed software. The computer controls the setting of the four circles, the filters, the shutters and the counter/timer.

If a crystal of unknown orientation is positioned at the centre of the instrument, it is oriented and its unit cell parameters are determined by the following procedures.

(a) Peak hunting – a defined area of reciprocal space is systematically examined by the peak hunting program. The process may take about two hours and is stopped when,

1. the whole area has been investigated,
2. a pre-set number of peaks have been found,
   or
3. the operator interrupts the program.

The 2θ, χ and ϕ angles together with the intensity are recorded for each peak.

(b) Determination of a primitive cell. In reciprocal space – this is carried out automatically by the computer using the data collected in (a). Each peak can be described in reciprocal space as a point

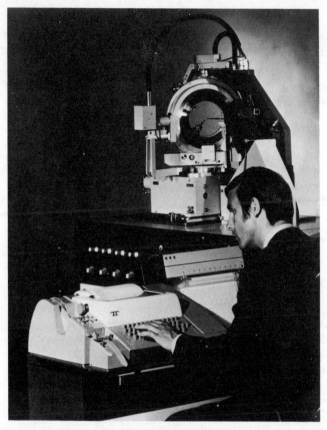

Figure 4.5. A Philips four-circle diffractometer

with co-ordinates x, y, z where

$$x = 2 \sin \theta \cos \chi \sin \phi$$
$$y = 2 \sin \theta \cos \chi \cos \phi$$
$$z = 2 \sin \theta \sin \chi$$

Each pair of points forms a vector and the computer finds the shortest vector (a^*), then the shortest vector not parallel to a^* ($= b^*$) and finally the shortest vector not in the a^*b^* plane ($= c^*$).

The three vectors describe a primitive cell and form the preliminary orientation matrix UB which is printed out.

$$UB = \begin{pmatrix} a_x^* & b_x^* & c_x^* \\ a_y^* & b_y^* & c_y^* \\ a_z^* & b_z^* & c_z^* \end{pmatrix}$$

A second matrix is also printed

$$M = \tilde{U}B \cdot UB \ (\tilde{U}B \text{ is the transpose of } UB)$$

The results obtained so far may now be interpreted by the crystallographer, and then a printout of the direct lattice constants is calculated by the computer from the following relationships and printed out:

$$UB^{-1} = \begin{pmatrix} a_x & a_y & a_z \\ b_x & b_y & b_z \\ c_x & c_y & c_z \end{pmatrix} \text{ and } M^{-1} = UB^{-1} \cdot \tilde{U}B^{-1}.$$

$$a = \lambda \sqrt{(M^{-1})_{11}}, \quad b = \lambda \sqrt{(M^{-1})_{22}}, \quad c = \lambda \sqrt{(M^{-1})_{33}}$$

$$\cos \alpha = \frac{\lambda^2 (M^{-1})_{23}}{bc}, \quad \cos \beta = \frac{\lambda^2 (M^{-1})_{13}}{ac}, \quad \cos \gamma = \frac{\lambda^2 (M^{-1})_{12}}{ab}$$

(c) Assignment of indices to the peaks and the refinement of UB—the computer calculates the Miller indices h, k, and l for all reflections on the list:

$$\begin{pmatrix} h \\ k \\ l \end{pmatrix} = UB^{-1} \begin{pmatrix} x \\ y \\ z \end{pmatrix}$$

and prints out hkl, xyz, d^{*2}, and the intensity. A least squares refinement program then refines the UB matrix and prints out the new UB matrix. This procedure may be repeated if required.

If the lattice parameters had previously been determined from, for example, film methods, then they can be supplied to the computer and UB may be adapted.

(d) Accurate determination of lattice constants—a program is used which investigates rows in the reciprocal lattice through the origin. The angles ω and 2θ are varied and χ and ϕ are fixed. The input needed by the program is:

1. the number of rows to be investigated, and
2. the direction of the rows given by their first reflection.

All the reflections in a row are measured and printed; the strong reflections are used in an accurate determination of the reciprocal lattice vectors.

Data Collection

The following input parameters are required before data collection begins:

1. Reflection measurement parameters: monochromator, balanced filters or β filter only; scanning speed; scan width; type of scan, ω, $\theta/2\theta$, or mixed; background measuring time; choice of scanning modes.
2. Input parameters defining the area to be measured.
3. Input parameters defining the sequence of measurement — allowance is made here for systematic absences in the diffraction data.
4. Reference reflections — a maximum of three reflections may be measured at given time intervals as a check that the quality of the data is not deteriorating.
5. Data collection program: this program controls the measurement and print out of data. An iterative procedure is followed: the indices of a reflection are calculated together with the necessary setting angles; the type of scan is initiated and various checks are carried out. Finally the data are printed and/or punched out.

   In addition to the procedures described above there are a number of sub-programs which allow the operation of the instrument to be independent of the main procedure.

The Siemens Hoppe Diffractometer (AED)

The on-line version of this diffractometer is available in three configurations, each of which uses a different computer for the control of the instrument. A Kristalloflex 4 (2 kW) generator is supplied together with the four-circle diffractometer and counting equipment, and a choice of computer may be made from:

1. 8K Siemens process computer 301
2. 8K Siemens process computer 302, 303, 304, or 305
3. 8K Digital Equipment Corporation PDP–8/I or PDP–8/L computer.

With all but the Siemens 301 computer it is possible to switch to off-line operation.

The following programs are supplied for the computers, enabling the precise position of coarsely pre-oriented crystal axes to be found, and a procedure to be followed for the automatic collection of diffraction data.

(a) Automatic control program – this program uses the unit-cell parameters to calculate the setting angles θ, χ and ϕ for each reflection. It sets the count-time and inserts an attenuation filter when necessary. The actual intensity count is made using a five-value measurement (see Figure 4.6) the sequence of which is also controlled

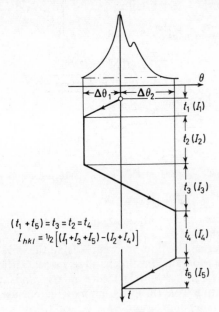

$$(t_1 + t_5) = t_3 = t_2 = t_4$$
$$I_{hkl} = \tfrac{1}{2}\left[(I_1 + I_3 + I_5) - (I_2 + I_4)\right]$$

Figure 4.6. Intensity determination according to the five-values method

by the program. A built-in check is made on the background counts each side of the peak to ensure the peak is actually at the calculated position. Check reflections are also made periodically under computer control. h, k, and l values are calculated for each reflection and the output of measured intensity data is obtained on punched paper tape and teleprinter print-out.

(b) Manual control program – the crystal may be oriented using this program and particular angular settings of the instrument may be selected. Counter measurements may be made at various points or a differential measurement involving step scans of varying number and step-size may be carried out. The insertion of either a K_β filter, balanced filters, or attenuation filters may also be effected by this program. The $\theta/2\theta$ clutch may also be operated.

(c) Reflection search program — the exact setting angles θ, χ and ϕ are calculated from the peak maxima of a number of reflections.

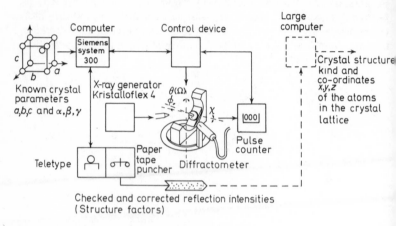

Figure 4.7. Automatic single-crystal diffractometer in on-line operation

The h, k, l indices and approximate θ, χ and ϕ values are needed as input for the program.

(d) Parameter calculation program — using the θ, χ, and ϕ values from the above program, the direct and reciprocal lattice parameters are calculated together with a lattice matrix and an adjustment matrix. The output from this program may then be used as input for the automatic control program.

Figure 4.7 shows the flow diagram for the operation of the on-line Siemens Hoppe Automatic Single Crystal Diffractometer.

The Stoe Four-Circle Diffractometer System STADI 4.

The STADI 4 system consists of the Stoe four-circle diffractometer with full circle Eulerian cradle of 336 mm diameter, including an optical alignment device (see Figure 4.8); the Stoe interface; a Digital Equipment Corporation PDP-8/E computer; a teletype; and a choice of various accessories. The complete system also contains a stabilised X-ray generator, and counting chain, and strip chart recorder.

Prior to the collection of intensity data the space group and unit cell parameters of the crystal under investigation should be known from Weissenberg and precession photographs. The crystal is then centred on the diffractometer by means of the microscope-telescope system and accurately re-aligned using either the four-circles of the diffractometer in a manual mode, or using the angular positions of two reflections of known Miller indices, and calculating an orientation matrix using the software package. An automatic routine also allows the determination of angular positions of

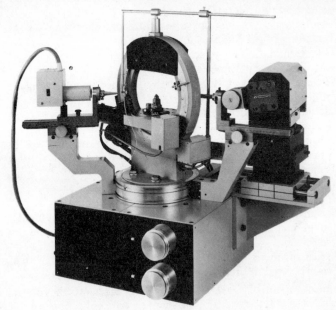

Figure 4.8. A Stoe STADI 4 diffractometer

reflection maxima to an accuracy of 0·01°. A least squares refinement procedure is used on a suitable number of reflections to calculate accurate lattice parameters.

The *software package* for STADI 4 comes in two parts, load 1 and load 2. Load 1 enables the crystal to be oriented and an orientation matrix to be determined. For each value of *hkl* the four angular values are calculated. A step by step improvement of the lattice parameters and the orientation matrix is made by comparing calculated and observed angular values. A more elegant procedure is also included in load 1 which enables the lattice parameters to be obtained from an automatic search for the positions of a number of reflection maxima followed by a least-squares procedure. Once the orientation matrix and lattice parameters have been determined sufficiently accurately, automatic data collection is carried out using load 2 of the software package.

The control of load 1 and load 2 depends on various console commands which are fed to the computer via the console of the teletype.

The following operations are executed through the console commands of load 1:

Input of X-ray wavelength and unit cell parameters.
Calculation of orientation matrix.
Calculation of the angular values for each reflection, h, k, l.
The setting of the required angular values.
The determination of reflection maximum.
Output of angular values.
Measurement step by step through a defined angular range.
Insertion of attenuation filters.
Calculation of lattice parameters by least squares.

Load 2 enables:

Various automatic modes of operation to be selected using console commands.
Measurement of integrated intensities.
Measurement of reflection profiles step by step.
Measurements with and without balanced filters.
Measurements with and without background measurements.
Measurements with and without automatic filter and speed selection.
Choice of up to three standard reference reflections which are re-measured at predetermined intervals.

The latest version of STADI 4 uses load 2 to accelerate the rotation of the circles so that the overall time spent on data collection is reduced.

The output from the STADI 4 system consists of the indices of the reflections hkl, the associated angular values, the number of counts and the time spent on counting. These values are both printed, and punched out on paper tape. Programs in FORTRAN IV are supplied to apply Lorentz/polarisation and absorption corrections to the measured data using a suitable computer.

Hilger & Watts Y290/FA328 Diffractometer

This instrument is a computer-controlled four-circle diffractometer which provides a fully automatic method of collecting and recording crystal data in a form suitable for further processing by a computer. Figure 4.9 illustrates the closed-loop system of operation that is used. The instrument (see Figure 4.10) comprises a four-circle diffractometer with direct positioning using a moiré fringe technique; an operator console which consists of a small computer (PDP-8/1) with its input/output device; and a 3kW X-ray generator

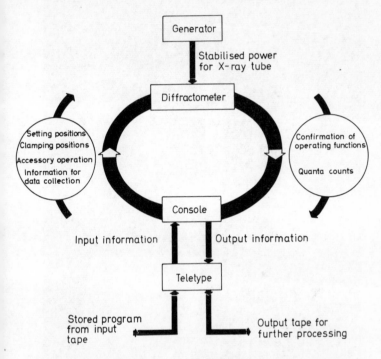

Figure 4.9. Closed loop system of operation. Input information is fed into the computer and circulates through the diffractometer. The quanta count is fed to the output device, and other information is recirculated through the computer

which takes a Philips tube type PW2113/00 (copper target with fine focus) or type PW2075/62 (molybdenum target with fine focus).

The *software* included with the system controls the instrument and contains all the instructions necessary to operate the diffractometer. It can be considered under three headings:

1. Background calculation routines
2. Control routines.
3. Supervisor routines.

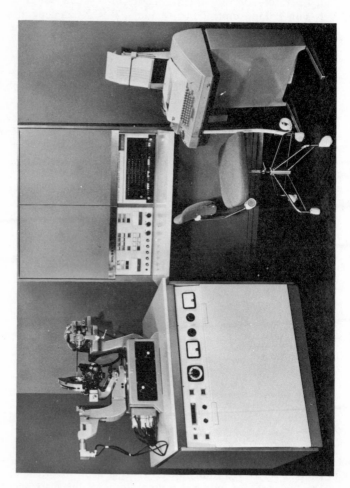

Figure 4.10. Hilger & Watts four-circle diffractometer

Background calculation routines—these include the initial angle calculation to bring the crystal into a reflecting position, and the calculation of the angular settings for each reflection so that the diffracted beam will lie in a plane common to the incident beam and the detector.

Control routines—these effectively control the state of the experiment by clamping the servo-mechanisms at set positions, counting Moiré fringes and operating the accessories such as filters, attenuators etc. The control program also includes the routine for counting X-ray quanta against real time.

Supervisor routines—these routines link the background and control routines to form an automatic system. They control the input/output devices for:

1. Intensity data.
2. The insertion of specific sub-routines.

The sequence of events performed by the system is as follows:

1. Determination of the next reflection *hkl*.
2. Calculation of the setting angles for this reflection.
3. The setting of these angles.
4. Counting the reflection.
5. Output.

In this sequence, (4) is the dominating factor although (3) (4) and (5) must occur sequentially. Time is saved by time-sharing these operations with:

1. The calculation of the angles for the next setting, and
2. The output of data from the previous position.

The system is always in manual mode on switching on; automatic operation is brought about by loading the necessary instructions into the computer via keyboard or paper tape.

The software package has two main programs, load 1 and load 2; the first allows semi-automatic (i.e. operator controlled) operation and the second is used for fully automatic data collection. There are additional extension programs that may be used in conjunction with load 1 and load 2, which make allowance for systematic absences in the diffraction data, allow various print-outs to be made, carry out least square refinement of crystal orientation and parameters, allow certain reflections to be specified for measurement, control the use of attenuators, and calculate lattice parameters from an orientation matrix as a quick alternative to the least squares calculation. In all there are 23 sub-routines provided.

The software package supplied with the Y290/FA328 system has recently been rewritten to take advantage of a 32K disc backing store, (FA 8238), that may be used with the basic 4K PDP8-series computer. This has allowed a major extension in the supervisor capability of the software giving advantages of a higher degree of automation of the system and greater ease and flexibility of use. (This rewritten software is only supplied with the FA 8238. The original unmodified software is supplied with the basic Y290/ FA328.)

Specific routines (supervisors) are called into the processor by typing a code of the form: *XY* on the Teletype. The routines are designated major or minor according to the time scale of the supervisor. Minor routines are exemplified by a request to drive the diffractometer to a given set of four circle angles. The supervisor completes its job in a few seconds. The major routines are exemplified by 'Peakfinder', in which the operator answers a series of questions posed by the computer in English language, e.g. wavelength, scan details and a list of reflections to be investigated. Peakfinder then determines the precise angular position of each reflection by step scanning, and uses this information to deduce the crystal orientation and lattice parameters by least square methods. The time scale of the Peakfinder supervisor is of the order of hours rather than seconds. No operator intervention or tape loading is needed after the operator has completed the initial dialogue.

Further examples of major supervisors are those controlling Automatic Data Collection and Initial Reflection Search. A.D.C. allows the measurement of integrated intensities from a crystal, and incorporates features such as Friedel pair measurement and the ability to correct for movement of the specimen during the experiment, without operator intervention. I.R.S. finds reflections from a crystal mounted in an unknown orientation, again without operator intervention.

Part 3
Crystal Structure Analysis—Treatment of the Measured Intensity Data

5 Overcoming the Phase Problem

5 Overcoming the Phase Problem

INTRODUCTION

The raw data which have been obtained by either photographic or counter methods are not in a suitable form for immediate use to determine a crystal structure. Various corrections must be made to the data such as Lorentz-polarisation corrections and absorption corrections, and for counter data allowance must also be made for the background counts, etc. The treatment of the data after correction then falls into two main categories depending upon the method to be used in solving the structure, i.e. indirect or direct methods. One common procedure, the *heavy-atom method*, is applicable where a heavy-atom in the crystal structure can be used to allocate an initial set of phases to the structure factors. The intensity of a reflection may be represented by $I (\propto F^2)$. $|F|$ is the structure amplitude and F is the structure factor. In a crystal belonging to a centrosymmetric space group, $\sqrt{F^2}$ can be either $+|F|$ or $-|F|$, depending largely upon the position of the heavy-atom whose contribution to the structure factor swamps the contributions from lighter atoms.

Once the phases of all the structure factors are known, a three-dimensional electron density map can be calculated which enables the positions of all the atoms in the unit cell to be found.

Direct methods are used usually where there are no heavy atoms in the structure, and they differ from the heavy-atom method in being completely objective and dependent solely on mathematical relationships and probabilities. The solution by direct methods can yield either a set of several possible electron density maps, when the most chemically reasonable one is chosen, or alternatively a single result may be obtained which can be right or wrong.

Whichever procedure is to be followed, it is at this point that use can be made of a computer. The most efficient system is one where a complete set of crystallographic programs can be stored on a tape or disk, and a separate tape or disk is used to store the data and results as they are obtained. In this way, once the data are on tape or disk it is only necessary to tell the computer which program and which data are to be used for a particular calculation. To work in this way, access is needed to a computer with a large store such as the KDF 9, Atlas, CDC 3600, or IBM 360.

The International Union of Crystallography publishes a list of crystallographic programs[37] describing the purpose of each, the language it is written in, and for which computer it has been written*. Whichever system is decided upon, heavy-atom or direct methods, the overall procedure is much the same. The data are converted into a form suitable for use by applying the necessary correction factors to the raw measured intensities. In the case of heavy-atom methods the corrected intensities are used to calculate either two projections or a three-dimensional Patterson synthesis in order to find the position of the heavy atom in the unit cell. The structure factors are then allocated phases using the heavy atom and a Fourier synthesis is calculated which gives a three-dimensional electron density map on which the other lighter atoms should be seen. A least-squares refinement program may be used to obtain the best fit of calculated and observed structure factors to the atomic parameters, and an overall scale factor. Once all the atomic positions have been satisfactorily allocated, a program to calculate inter-atomic distances and angles may be run. If a graph plotter is available, a program such as ORTEP[38] can be used to draw the atoms as ellipsoids of vibration seen from any chosen direction.

There are many other crystallographic programs; peak searches can be carried out in the Patterson and Fourier maps, H-bond searches can be made between stated limits of angles and distances. H atom co-ordinates can be calculated, and corrections to atomic co-ordinates can be made to allow for rigid body vibrations by the molecule.

In the case of direct methods, a similar approach may be used; the required programs are called from the magnetic tape or disk, and the corrected data are treated accordingly. Once the structure factors have been allocated phases, Fourier syntheses and least-squares refinement cycles are carried out as in the heavy-atom method. Before looking in more detail at the programs, it is necessary to consider some of the expressions used in the calculations, and the definition of some of the terminology that is used.

Consider the following flow diagram which shows the procedure that may be followed when either heavy-atom or direct methods are used in crystal structure analysis. Each of the steps that are taken using the heavy atom method will then be considered in turn, and finally the application of direct methods will be considered separately.

*See Chapter 6.

Measured raw intensities $I(hkl)$

↓

The application of corrections to the measured data;
data reduction.

Heavy atom method   *Direct methods*

Calculation of a Patterson synthesis   Calculation of normalised
structure factors (E_h)

Calculation of a Fourier synthesis   Calculation of the signs of
E_h, using mathematical
relationships

Calculation of an E-map
(Fourier Synthesis)

Least squares refinement

Final difference Fourier
calculation

CORRECTIONS APPLIED TO THE DATA

The Lorentz Factor, L

The Lorentz factor is an expression for the time a plane of a rotating
crystal spends in the reflecting position. In terms of the reciprocal
lattice/reflecting sphere concept, this is the length of time the lattice
point is in contact with the sphere of reflection, and this is obviously
dependent on the distance of the reciprocal lattice point from the
origin. At large Bragg angles (θ) the reciprocal lattice point passes
almost tangentially through the sphere of reflection, and L is large.
At low Bragg angles, L is again large, as the point is near the origin
of the reciprocal lattice and the time taken in cutting the sphere of
reflection is large. L is a minimum at $\theta = 45°$. The expression for L
depends upon the diffraction geometry used during the collection
of intensity data. In the general case the following expression can be
derived for L:

$$L = 1/\cos \mu \cos v \sin \gamma$$

where μ, v, and γ are the diffractometer setting angles mentioned earlier. This expression can be greatly simplified for normal-beam equatorial geometry when:

$$L = 1/\sin 2\theta, \quad \text{as } \mu = v = 90° \quad \text{and} \quad \gamma = 2\theta$$

In the equi-inclination case $\mu = -\mu$ and $L = 1/\cos^2 \mu \sin \gamma$.

The Polarisation Factor, *p*

When X-rays are reflected by a plane in a crystal, a reduction in the intensity of the reflected beam occurs, as a result of polarisation. The polarisation factor p is defined as $\frac{1}{2}(1 + \cos^2 2\theta)$.

If $\theta = 0$ or $90°$, $p = 1$ i.e. there is no polarisation.

When $\theta = 45°$, $p = \frac{1}{2}$ and the reflected beam is completely polarised.

In terms of the angles μ, v, and γ, the following expressions can be derived for the polarisation factor:

Normal beam: $\mu = 0$, $p = \frac{1}{2} + \frac{1}{2}\cos^2 v \cos^2 \gamma$

Equi-inclination: $\mu = -v$, $p = \frac{1}{2} + \frac{1}{2}\sin^4 v - \frac{1}{4}\sin^2 2v \cos \gamma + \frac{1}{2}\cos^4 v \cos^2 \gamma$

Discussions of Lorentz and polarisation factors are given in detail in, for example, *Crystal Structure Analysis*, M. J. Buerger, p. 156–186; *Single Crystal Diffractometry*, V. W. Arndt and B. T. M. Willis, p. 278–288; and *International Tables* Vol. II, p. 265.

Usually the corrections for Lorentz and polarisation factors (Lp) are combined and applied together. For equi-inclination Weissenberg photographs:

$$Lp^{-1} = \frac{2\cos^2 \mu \sin \gamma}{1 + \cos^2 \theta}$$

Tables of $2Lp$ as a function of $\sin \theta$ are given in *International Tables* Vol. II, p. 268.

The Philips Spot Shape Correction[39]

It is characteristic of higher layer equi-inclination Weissenberg photographs that spots on one half of the film are compacted, and on the other half are extended. It is usual to measure the intensities of the extended spots and apply a Philips spot shape correction to the values obtained,

i.e. $I = WI_e$ where I = the corrected intensity and I_e = the measured extended spot intensity.

$W = (A + \Delta A)/A =$ the fractional area increase of an extended spot

$$= 1 + \frac{180}{4\pi} \cdot \frac{\zeta(\xi_m^2/\xi^2 - 1)^{\frac{1}{2}}}{R_1/(1 - \zeta^2/4)^{\frac{1}{2}} + R_2}$$

ξ_m is the maximum observable value of $\xi(= 4 - \zeta^2)^{\frac{1}{2}}$, R_1 and R_2 are instrumental constants, $R_1 =$ camera radius and $R_2 =$ X-ray source pinhole to crystal distance. In most commercial instruments $R_1 = 28\cdot7$ mm, and $R_2 = 75\cdot0$ mm.

The Absorption Correction

The measured intensities obtained by X-ray diffraction from single crystals often need to be corrected for absorption of radiation by the specimen. Absorption effects may be noticed when intensities of symmetry related reflections, which have been corrected for other

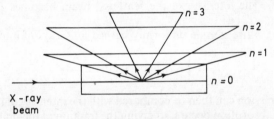

Figure 5.1. Absorption by a platey crystal. Diffracted rays lying on the cone n = 0 have the greatest crystal path length, and those on n = 3 have the least path length

effects, are compared and found to differ. The effects of absorption often manifest themselves in the refinement of a structure when physically meaningless values may be obtained for the vibration parameters of the atoms. For example, isotropic atomic vibration parameters may appear with negative values.

The general effect can be seen if a platey crystal is considered, when data are collected with the plate parallel to the X-ray beam (see Figure 5.1). The paths of the zero layer reflections will all lie completely in the crystal plate. However, higher layer diffraction will progressively travel through less and less of the crystal, and the diffracted rays will suffer less attenuation the higher the layer. This effect will be more noticeable in strongly absorbing material, and for this reason it is more important to correct for the effects of absorption in metal complexes than in organic crystals.

Busing and Levy[40] describe a method of applying absorption corrections to zero level reflections that depends upon the availability of a high speed computer, and Wells[41] describes an extension to their work that has been applied to Weissenberg and precession cameras, which also allows a correction to be made when the sample is mounted in an absorbing cylindrical container, possibly with absorbing liquid trapped between the sample and the tube.

If an incident X-ray beam of intensity I_0 falls on absorbing material, then at a depth t the intensity will be given by:
$I_t = I_0 \exp(-\mu t)$ where μ is the linear absorption coefficient of the material

If the volume element causing the scattering is dV then the scattered intensity will be proportional to dV and I_t, i.e.,

$I_s = K I_t dV$ where I_s is the intensity of the scattered beam
$I_s = K I_0 \exp(-\mu t)$

The scattered ray will then travel a further distance t' before emerging from the absorber, and will undergo further attenuation by a factor $\exp(-\mu t)$. The intensity of the scattered beam becomes $I_s = K I_0 \exp[-\mu(t+t')] dV$.

For the entire sample uniformly bathed in X-rays of intensity I_0,

$$I_s = \int K I_0 \exp[-\mu(t+t')] \, dV$$

This expression can then be compared with the intensity of the beam when no absorption occurs, so giving the fraction of the unabsorbed beam that would be measured on an intensity film:

$$A = \frac{1}{V} \exp[-\mu(t+t')] \, dV$$

The integration is over the volume of the crystal V, t is the primary beam path length and t' the diffracted beam path length. A is the absorption correction factor.

Busing and Levy[40] describe a method of evaluating this expression, and for the purpose of the calculation, each face of the crystal is described by the intercepts it makes on a set of orthogonal axes by the expression $ax + by + cz - d = 0$. The orthogonal axes x, y, z have a known relationship to the crystal axes a, b, c, and it is assumed that the crystal has no re-entrant angles.

Wells[41] describes the arguments that lead to the determination of the direction cosines of the incident and reflected rays for different camera geometries.

It is often sufficient to apply an approximate absorption correction to intensity data, and if the crystal can be approximated to a sphere or a cylinder absorption corrections can be made using the tabulated values of correction factors given by Evans and Ekstein[42] and Bradley[43] respectively.

Extinction[44]

Extinction, which is attenuation of the primary beam as it passes through a crystal and should not be confused with systematic extinction, which is the absence of certain reflections due to space group symmetry, was first investigated by Darwin[45], who divided the phenomenon into two types, primary and secondary. Lonsdale[46] has also discussed the subject.

Primary extinction arises from destructive interference with the primary beam by secondary beams that have been multiply reflected by the crystal planes. Figure 5.2 shows how this arises.

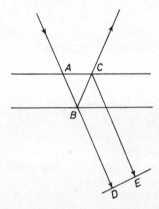

Figure 5.2. Primary extinction

The ray at A produces a Bragg reflection in the direction BC. This ray is at the correct angle to be reflected again at C back in the direction of the primary beam. As reflection is accompanied by a phase change of $\pi/2$, then the ray at E which has been reflected twice will be out of phase with the ray at D by π, and destructive interference will weaken the primary beam D as it continues through the crystal. Progressive attenuation of the primary beam will occur as it passes through the crystal, as a result of this type of multiple reflection.

The effect is most noticeable for strong reflections at low θ values, i.e. reflections occurring near the equator of zero layer Weissenberg films. During the refinement of the structure the magnitude of the observed structure factor (F_0) for such reflections will appear much less than that of the calculated one (F_c).

Secondary extinction again reduces the intensity of the primary beam, but in this case it arises where some reflections are so intense that they attenuate the primary beam as it passes through the crystal. Lower level planes in the crystal thus receive a less intense primary beam than higher level ones. Generally this effect is observed at low θ angles where the stronger reflections usually occur.

This effect is similar to absorption, and Darwin has shown how allowance may be made for it in the linear absorption coefficient of an absorption correction.

Sometimes extinction effects may be overcome by quenching the crystal in liquid air. The phenomenon of extinction is a well known effect which is difficult to allow for, and in most cases where it is manifest the affected data are usually removed.

Zachariasen[47] describes a procedure for treating secondary extinction, and Asbrink and Werner[48] discuss its application to an irregular crystal.

HEAVY ATOM METHOD

Patterson Synthesis

The intensities of the diffracted rays that are measured when data are collected during a structure determination do not provide all the information that is needed to solve the structure. The structure amplitude $|F|$ is readily obtained, but the phase associated with it is not known. As explained elsewhere, for a centrosymmetric structure this means that the sign ($+$ or $-$) of the structure factor cannot be determined. If both the magnitude and the phase of the structure factor were known for all the hkl reflections, then an electron density map of the unit cell could be obtained from a Fourier synthesis which is described in the next section.

In an attempt to overcome this phase problem Patterson[49, 50] used the squares of the structure amplitudes, $|F|^2$, as Fourier coefficients. He defined a function:

$$P(UVW) = V \int_0^1 \int_0^1 \int_0^1 \rho(xyz)\rho(x+U, y+V, z+W)\,dxdydz \qquad (5.1)$$

where $\rho(xyz)$ is the electron density at the point x, y, z. The Fourier expression for the electron distribution is:

$$\rho(xyz) = \frac{1}{V} \sum_h \sum_k \sum_l F(hkl) \exp\left[-2\pi i(hx+ky+lz)\right] \qquad (5.2)$$

where h, k, l the indices of the reflecting planes are summed from $-\infty$ to $+\infty$.

From Eq. (5.1) and (5.2) it can be shown that:

$$P(UVW) = \frac{1}{V} \sum_h \sum_k \sum_l |F(hkl)|^2 \exp[2\pi i(hU + kV + lW)]$$

This is the expression known as the Patterson function*, and a peak occurring at the point $P(UVW)$ where U, V and W are the co-ordinates of the point P on a Patterson vector map results from atoms in the unit cell at points x, y, z; and x', y', z'; where

$$U = x - x',$$
$$V = y - y', \text{ and}$$
$$W = z - z'$$

If one compares the co-ordinates of two identical atoms in adjacent unit cells, it will be seen that $x = x'$, $y = y'$, and $z = z'$, which leads to a peak at the origin where $U = 0$, $V = 0$, and $W = 0$. Every atom therefore contributes to the origin peak on the Patterson vector map. If a projection down the a-axis is calulated then the above expression reduces to:

$$P(VW) = \frac{1}{A} \sum_k \sum_l |F(0kl)|^2 \exp[2\pi i(kV + lW)]$$

All the information for such a calculation will be contained on the a-axis zero layer Weissenberg photograph. To calculate a Patterson projection down any unit cell axis, it is only necessary to have intensity data from the corresponding zero layer Weissenberg photograph.

Characteristics of a Patterson Vector Map[51]

A three-dimensional Patterson synthesis provides a vector map of the contents of the unit cell of the crystal. The value of $P(UVW)$ will be zero everywhere except where the values of UVW represent a vector between two atoms. In certain circumstances, which are

---

*The quantity $P(UVW)$ is real for all values of U, V, W and the form of the expression that is usually used is:

$$P(UVW) = \frac{1}{V} \sum_h \sum_k \sum_l |F(hkl)|^2 \cos 2\pi(hU + kV + lW)$$

described in the following sections, some or all of the atomic co-ordinates of the crystal structure can be determined from such a map. The following characteristics are displayed by a Patterson vector map:

1. Every pair of atoms in the unit cell will produce a peak on the vector map, therefore if there are N atoms in the unit cell there will be N^2 peaks in the vector map.
2. N of these peaks will correspond to the vector between each atom and itself, and these occur at the origin of the vector map. There are, therefore, $N^2 - N$ non-origin peaks, i.e. $N(N-1)$.
3. The symmetry of a crystal's unit cell can be described by one of 230 space groups. Only 24 space groups are needed to des-cribe the possible symmetry of a vector map. Buerger[52] gives three theorems that can be used to derive the vector space group of a crystal from its real space group. and gives a table showing every space group and its vector equivalent.

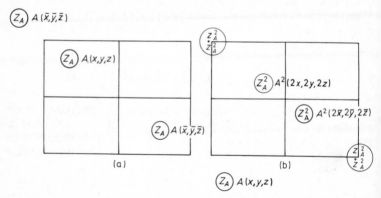

Figure 5.3. Real space and vector space. (a) A unit cell space group P1̄ containing two atoms of atomic weight Z_A. (b) Patterson vector map produced by (a)

4. Each peak on a vector map corresponding to two atoms in the unit cell of atomic numbers Z_1 and Z_2, will have $Z_1 . Z_2$ electrons contained in its volume. The volume of the vector peak is related to the density of the atoms it represents; therefore care must be taken in interpreting vector maps from peak height alone.
5. The Patterson function is centrosymmetric, so every peak on the vector map will be related to a similar peak by a centre of symmetry.

If a Patterson synthesis is calculated for a crystal containing a heavy atom, then maxima on the vector map which represent heavy atom–heavy atom vectors, will appear as much larger peaks than any of the others. Other large peaks may also be present due to the overlap of smaller peaks, but this is more of a problem in Patterson projections than in three-dimensional Patterson syntheses.

If we consider the space group $P\bar{1}$ as an example, and take the case where there are two asymmetric units per unit cell, the co-ordinates of the equivalent positions are x, y, z; $\bar{x}, \bar{y}, \bar{z}$. If each asymmetric unit contains a heavy atom then the co-ordinates of the maxima in the vector map which represents a vector between the heavy atoms will be $(x, y, z) - (\bar{x}, \bar{y}, \bar{z}) = 2x, 2y, 2z$, i.e. $U = 2x$, $V = 2y$, $W = 2z$. The heavy atom co-ordinates in the unit cell will then be half the co-ordinates of the largest peak on the Patterson map. Figure 5.3 shows the relationship between such a unit cell and its vector-space equivalent.

Special sections of a vector map, known as Harker sections, contain information about atoms that are related by certain symmetry operations in the unit cell. These will now be considered.

Harker Sections

It has been shown[53] that if a crystal contains axes or planes of symmetry, then certain planes or lines in the vector representation of the unit cell will contain useful information that leads to the determination of some of the atomic co-ordinates in the unit cell. For example[54]:

1. If the unit cell contains a two-fold axis parallel to the b-axis of the unit cell, then for every atom at x, y, z, there will be a symmetry related atom at $\bar{x}, y, \bar{z}$. A maximum will occur on the vector map at $U = 2x$, $V = 0$, $W = 2z$ (i.e. the difference between the two sets of co-ordinates). A vector-space section taken perpendicular to the b-axis at $V = 0$ will contain all such maxima. From the co-ordinates of the maxima the x and z co-ordinates of the atoms in the unit cell can be obtained simply by halving their values.

2. In the case of a two-fold screw-axis parallel to the b-axis, equivalent atoms will have co-ordinates x, y, z; and $\bar{x}, y+\frac{1}{2}, \bar{z}$. The corresponding vector will have a maximum at $U = 2x$, $V = \frac{1}{2}$, $W = 2z$ in vector space. The vector distances of the atoms from the screw-axis therefore can be found on this section.

3. If a mirror plane lies in the crystal perpendicular to the b-axis then atoms related by the plane will produce maxima along the b-axis of the vector map, i.e. the line $U = 0$, $V = y$, $W = 0$ and the distance of the atoms from the plane can be found.

4. A glide-plane perpendicular to the b-axis with a translation $c/2$ will result in vector-space maxima along the line $U = 0$, $V = y$, $W = \frac{1}{2}$.

Simplification of the Patterson Function

The Patterson function may be simplified to allow for the symmetry relationships that exist in the different crystal systems between the various values of the observed $|F(hkl)|$.

The following nine expressions are obtained. (Further simplification is in fact possible. See for example D. Harker, *J. chem. Phys.* **4**, 381 (1936)).

Triclinic

$$P(UVW) = \frac{2}{V_c} \sum_0^\infty \sum_0^\infty \sum_0^\infty \{ |F(hkl)|^2 \cos 2\pi(hU + kV + lW) +$$

$$|F(\bar{h}kl)|^2 \cos 2\pi(-hU + kV + lW) +$$

$$|F(h\bar{k}l)|^2 \cos 2\pi(hU - kV + lW) +$$

$$|F(hk\bar{l})|^2 \cos 2\pi(hU + kV - lW)\}$$

Monoclinic

$$P(UVW) = \frac{4}{V_c} \sum_0^\infty \sum_0^\infty \sum_0^\infty \{ \overset{c \, \text{axis unique}}{|F(hkl)|^2} \cos 2\pi(hU + kV) +$$

$$|F(\bar{h}kl)|^2 \cos 2\pi(hU - kV)\} \cos 2\pi lW$$

$$P(UVW) = \frac{4}{V_c} \sum_0^\infty \sum_0^\infty \sum_0^\infty \{ \overset{b \, \text{axis unique}}{|F(hkl)|^2} \cos 2\pi(hU + lW) +$$

$$|F(\bar{h}kl)|^2 \cos 2\pi(hU - lW)\} \cos 2\pi kV$$

Orthorhombic

$$P(UVW) = \frac{8}{V_c} \sum_0^\infty \sum_0^\infty \sum_0^\infty |F(hkl)|^2 \cos 2\pi hU \cos 2\pi kV \cos 2\pi lW$$

Tetragonal

$$P(UVW) = \frac{4}{V_c} \sum_0^\infty \sum_0^\infty \sum_0^\infty \{ |F(hkl)|^2 \cos 2\pi(hU+kV) + \overset{4,\bar{4},4/m}{}$$

$$|F(\bar{h}kl)|^2 \cos 2\pi(hU-kV)\} \cos 2\pi lW$$

$$P(UVW) = \frac{8}{V_c} \sum_0^\infty \sum_0^\infty \sum_0^\infty |F(hkl)|^2 \cos 2\pi hU \overset{422,4mm,\bar{4}2m,4/mmm}{}$$

$$\cos 2\pi kV \cos 2\pi lW$$

Trigonal

$$P(UVW) = \frac{2}{V_c} \sum_0^\infty \sum_0^\infty \sum_0^\infty \{ |F(hkl)|^2 \overset{3,\bar{3},32,3m,\bar{3}m}{}$$

$$\cos 2\pi(hU+kV+lW) + |F(\bar{h}kl)|^2 \cos 2\pi(-hU+kV+$$

$$lW) + |F(h\bar{k}l)|^2 \cos 2\pi(hU-kV+lW) +$$

$$|F(hk\bar{l})|^2 \cos 2\pi(hU+kV-lW)\}$$

Hexagonal

$$P(UVW) = \frac{4}{V_c} \sum_0^\infty \sum_0^\infty \sum_0^\infty$$

$$\overset{6,\bar{6},6/m,622,6mm,\bar{6}m2,6/mmm}{}$$
$$\{ |F(hkl)|^2 \cos 2\pi(hU+kV) + |F(\bar{h}kl)|^2$$

$$\cos 2\pi(hU-kV)\} \cos 2\pi lW$$

Cubic

$$P(UVW) = \frac{8}{V_c} \sum_0^\infty \sum_0^\infty \sum_0^\infty$$

$$\overset{23,m3,432,\bar{4}3m,m3m}{}$$
$$|F(hkl)|^2 \cos 2\pi hU \cos 2\pi kV \cos 2\pi lW$$

Examples of the Determination of Atomic Co-ordinates from Patterson Syntheses

1. *The platinum atom co-ordinates of* $Pt(NH_3)_2Cl_2$ [12]

The peaks on a vector map that represent heavy atom–heavy atom vectors in real space, together with a knowledge of the co-ordinates of equivalent positions in the unit cell (which are given for every space group in *International Tables* Vol. I), may lead to an un-equivocal determination of the unit cell co-ordinates of the heavy atom. On the other hand, there may be more than one set of co-ordinates in the unit cell which could produce the peaks of the vector map.

Figures 5.4 and 5.5, show the Patterson projections and the corresponding unit cell projections down the a- and b-axes of the compound cis-$Pt(NH_3)_2Cl_2$. The crystal space group is $P\bar{1}$, and there are two molecules per unit cell. The co-ordinates of equivalent positions are x, y, z; $\bar{x}, \bar{y}, \bar{z}$. Subtracting these gives the expected

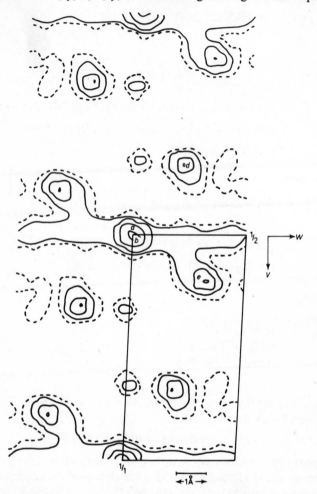

Figure 5.4(a). A Patterson projection down the a axis of cis-Pt(NH₃)₂Cl₂. The zero contour is shown by a dotted line

co-ordinates of the platinum atom–platinum atom maximum in vector space as $U = 2x$, $V = 2y$, $W = 2z$. This is illustrated in Figure 5.6; $\odot$ and $\odot$, represent symmetrically equivalent enantio-morphous positions. The signs $+$ and $-$ indicate that the position shown is a distance z up or down the vertical z-axis. It should be possible to obtain values of x, y, z, from the two Patterson projections. An examination of the projection down the a-axis at first glance shows a large peak only at the origin. Closer examination

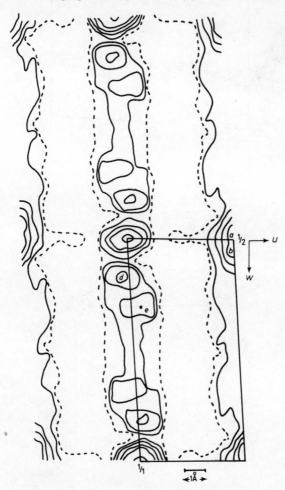

Figure 5.4(b). A Patterson projection down the b axis of cis-Pt(NH₃)₂Cl₂. The zero contour is shown by a dotted line.

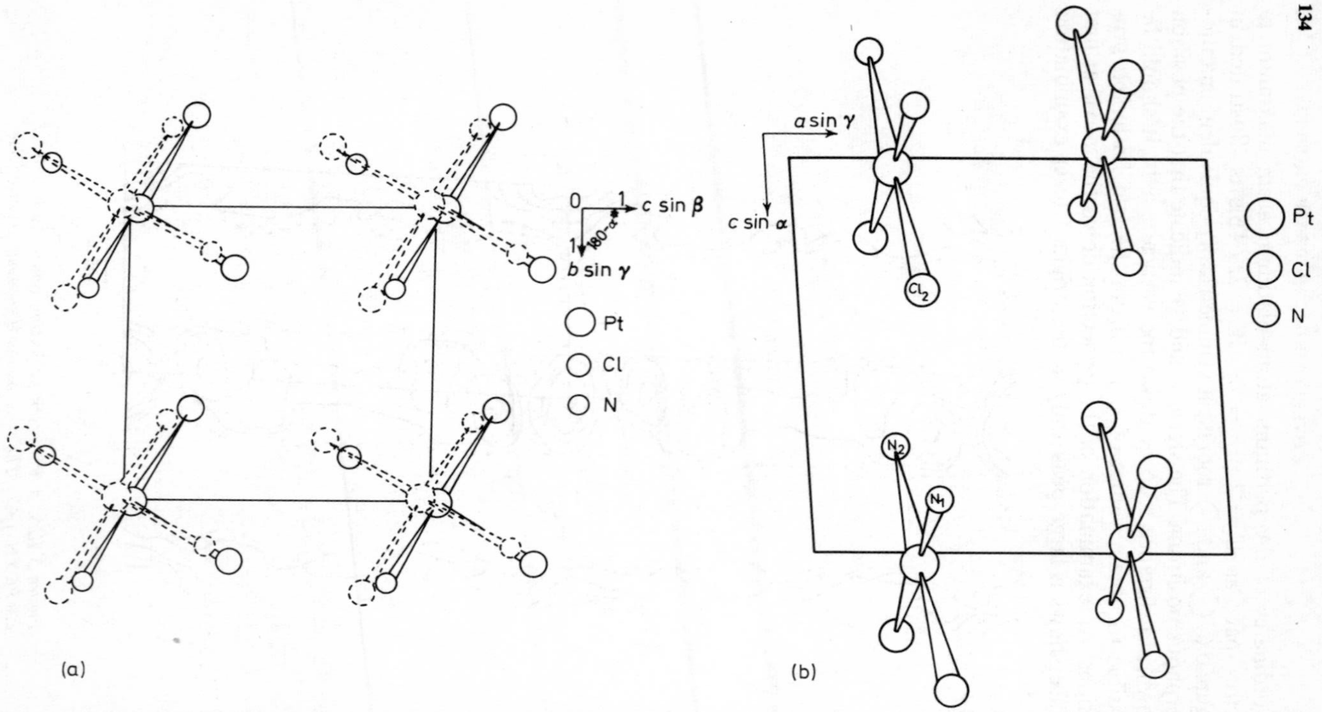

Figure 5.5. Projections down the a and b axes of the unit cell of cis-Pt(NH₃)₂Cl₂. (a) Projection down the a-axis. (b) Projection down the b-axis

of the contours reveals that this is in fact dumb-bell shaped and consists of two overlapping peaks. This is confirmed in the projection down the b-axis where a clearer separation of the peaks is seen halfway along the U direction, peaks a and b. The peaks d and e represent Pt–Cl vectors, and to obtain a unique set of co-ordinates for the Pt atom in the unit cell it was only necessary to show which of the two possible positions of the platinum atom gave the best agreement between observed and calculated structure factors. (In fact the vector space maximum at b was associated with the Pt–Cl maxima d and e.)

The Patterson projection down the a-axis gave the co-ordinates $U = 0, V = 2y, W = 2z$ of the Pt–Pt peak; the projection down the

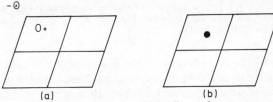

Figure 5.6. Spacegroup P1. (a) Real cell co-ordinates: x, y, z; $\bar{x}, \bar{y}, \bar{z}$. (b) Vector cell co-ordinates: U(= 2x), V(= 2y), W(= 2z)

b-axis gave the co-ordinates $U = 2x, V = 0, W = 2z$. In this way the x, y, z, co-ordinates of the Pt atom in the unit cell were found.

2. *The copper atom in a copper-glycyl-L-glutamic acid complex*[55]
The space group of the crystals is $C222_1$, which is non-centrosymmetric. There are 8 copper atoms per unit cell and the co-ordinates of equivalent positions are:

$$x, y, z; \quad x, \bar{y}, \bar{z}; \quad \bar{x}, \bar{y}, \tfrac{1}{2} + z; \quad \bar{x}, y, \tfrac{1}{2} - z$$

Four other positions exist, as a result of the C-face centring at $\tfrac{1}{2}$, $\tfrac{1}{2}$, 0; but for the present purpose these can be ignored.

The co-ordinates of the heavy-atom peaks in the Patterson synthesis can be obtained by subtracting the co-ordinates of each equivalent position in the unit cell from all the others. This is best illustrated by the following table and Figure 5.7:

	x, y, z	$x, \bar{y}, \bar{z}$	$\bar{x}, \bar{y}, \tfrac{1}{2}+z$	$\bar{x}, y, \tfrac{1}{2}-z$
x, y, z	0	$0, 2y, 2z$	$2x, 2y, -\tfrac{1}{2}$	$2x, 0, -\tfrac{1}{2}+2z$
$x, \bar{y}, \bar{z}$	$0, \bar{2}y, \bar{2}z$	0	$2x, 0, -\tfrac{1}{2}-2z$	$2x, 2y, -\tfrac{1}{2}$
$\bar{x}, \bar{y}, \tfrac{1}{2}+z$	$\bar{2}x, \bar{2}y, \tfrac{1}{2}$	$\bar{2}x, 0, \tfrac{1}{2}+2z$	0	$0, \bar{2}y, 2z$
$\bar{x}, y, \tfrac{1}{2}-z$	$\bar{2}x, 0, \tfrac{1}{2}-2z$	$\bar{2}x, 2y, \tfrac{1}{2}$	$0, \bar{2}y, 2z$	0

The diagonal of zeros in the table arises from the subtraction of like co-ordinates.

The signs '$\frac{1}{2}+$' and '$\frac{1}{2}-$' indicate the position is a distance $\frac{1}{2}+z$ or $\frac{1}{2}-z$ along the z-axis.

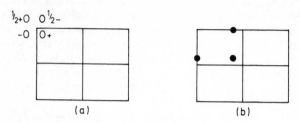

Figure 5.7. Space group $C222_1$. (a) Real cell co-ordinates: x, y, z; $x, \bar{y}, \bar{z}$; $\bar{x}, \bar{y}, \frac{1}{2}+z$; $\bar{x}, y, \frac{1}{2}-z$. (b) Vector cell co-ordinates: $U(=0)$, $V(=2y)$, $W(=\bar{2}z)$; $U(=2x)$, $V(=2y)$; $W(=-\frac{1}{2})$; $U(=2x)$, $V(=0)$, $W(=2z-\frac{1}{2})$

Which of the co-ordinates in the above table are used depends upon which part of the Patterson map has been calculated. It would be a waste of time and money to calculate a full three-dimensional synthesis, and it is usual to calculate only an asymmetric part of the three-dimensional map which will contain all the information needed. The whole cell vector map can be constructed from this asymmetric part if it is required.

When deciding which part to calculate it is necessary to consider the symmetry of the vector map. In the case of the space group $C222_1$ the vector map symmetry is *Cmmm*. If a volume represented by $U = 0$ to $U = \frac{1}{2}$, $V = 0$ to $V = \frac{1}{4}$, and $W = 0$ to $W = -\frac{1}{2}$ (the fractions represent the lengths of the cell edges of the vector map) is calculated, it should contain all the information needed to obtain values of x, y, z, the co-ordinates of the copper atom in the unit cell.

An examination of the table of vector space-co-ordinates shows that both above and below the diagonal there are three pairs of equivalent co-ordinates. Overall there are three groups of four equivalent sets of co-ordinates. In the section of vector space that has been calculated there should be three copper–copper maxima with co-ordinates $U = 0$, $V = 2y$, $W = \bar{2}z$; $U = 2x$, $V = 2y$, $W = -\frac{1}{2}$; $U = 2x$, $V = 0$, $W = -\frac{1}{2}+2z$. This was in fact the case, and from these peaks the x, y, z, co-ordinates of the copper atom in the unit cell were obtained.

3. *The copper atom in the sodium complex, sodium perchlorate* $-$ Bis NN'-ethylenebis(salicylideneiminato)-Cu(II)[56]
The systematic absences found on the Weissenberg film data showed that the space group of the crystals is either *Cc* or *C2/c*. The latter

was in fact the case. There are 8 Cu atoms per unit cell. Ignoring the C-face centring, the co-ordinates of the other 4 equivalent positions are x, y, z; $\bar{x}, \bar{y}, \bar{z}$; $\bar{x}, y, \frac{1}{2}-z$; $x, \bar{y}, \frac{1}{2}+z$.

The following table and Figure 5.8 give the vector-peak co-ordinates derived from the equivalent positions.

	x, y, z	$\bar{x}, \bar{y}, \bar{z}$	$\bar{x}, y, \frac{1}{2}-z$	$x, \bar{y}, \frac{1}{2}+z$
x, y, z	0	$2x, 2y, 2z$	$2x, 0, -\frac{1}{2}+2z$	$0, 2y, -\frac{1}{2}$
$\bar{x}, \bar{y}, \bar{z}$	$\bar{2}x, \bar{2}y, \bar{2}z$	0	$0, \bar{2}y, -\frac{1}{2}$	$\bar{2}x, 0, -\frac{1}{2}-2z$
$\bar{x}, y, \frac{1}{2}-z$	$\bar{2}x, 0, \frac{1}{2}-2z$	$0, 2y, \frac{1}{2}$	0	$\bar{2}x, 2y, \bar{2}z$
$x, \bar{y}, \frac{1}{2}+z$	$0, \bar{2}y, \frac{1}{2}$	$2x, 0, \frac{1}{2}+2z$	$2x, \bar{2}y, 2z$	0

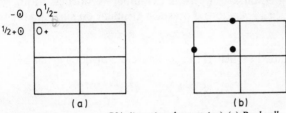

Figure 5.8. Space group C2/c (ignoring the centring). (a) Real cell co-ordinates: x, y, z; $\bar{x}, \bar{y}, \bar{z}$; $\bar{x}, y, \frac{1}{2}-z$; $x, \bar{y}, \frac{1}{2}+z$. (b) Vector cell co-ordinates: $U(= 2x)$, $V(= 2y)$, $W(= 2z)$; $U(= 0)$, $V(= 2y)$, $W(= \frac{1}{2})$; $U(= 2x)$, $V(= 0)$, $W(= \frac{1}{2}+2x)$

The x, y, z unit cell co-ordinates of the copper atom can be obtained from the vector-peak co-ordinates $U = 2x$, $V = 2y$, $W = 2z$; $U = 0$, $V = 2y$, $W = \frac{1}{2}$; and $U = 2x$, $V = 0$, $W = \frac{1}{2}+2z$.

In this structure there are four sodium atoms which lie on two-fold axes. Ignoring the two at the C-face centring, the co-ordinates of the equivalent positions of the other two sodium atoms are $0, y, \frac{1}{4}$; and $0, \bar{y}, \frac{3}{4}$. A sodium atom vector peak therefore appears at $U = 0$, $V = 2y$, $W = \frac{1}{2}$; on the vector map. A similar argument applies to the chlorine atoms in the molecule.

From what has been said so far on the interpretation of Patterson maps the reader may consider the procedure to be quite straightforward and wonder why so much is made of the phase problem. In fact the interpretation of Patterson maps is often far from straightforward and each case that is considered usually poses special problems of its own. Books have been written* which deal solely with vector space and the interpretation of vector maps.

There are many reasons why particular structures present problems, for example organic molecules often have no heavy atom

*For example *Vector Space*, M. J. Buerger, Wiley, New York (1959)

whose co-ordinates in the unit cell may be obtained from the vector map. In addition the interatomic vectors in organic molecules are often similar in length, which leads to peak overlap in many cases in the vector map. This latter problem also arises since the cell of the Patterson map and the unit cell have the same volume, but the Patterson cell is much more densely populated by peaks of greater volume.

Various methods have been devised for dealing with such problems, and one of the most popular is the *sharpened Patterson* with the origin removed. In a sharpened Patterson synthesis the atoms are treated as point atoms where the scattering power of the atom is a constant equal to Z its atomic number. A 'normal' atom, scattering factor f, has a scattering power dependent on $\sin \theta/\lambda$, and

$$f = f_0 \, e^{-B(\sin^2 \theta)/\lambda^2}$$

as explained earlier. If the structure factor, F, is given by

$$F = f \sum_{j=1}^{N} e^{2\pi i(hx + ky + lz)}$$

then a comparison of F_N, the structure factor of a normal atom and F_p that of a point atom, gives:

$$\frac{F_p}{F_N} = \frac{Z}{f_0 \, e^{-B(\sin^2 \theta)/\lambda^2}}$$

and

$$F_p = F_N \cdot Z/f_0 \, e^{-B(\sin^2 \theta)/\lambda^2}$$

Using this expression values of $|F_p|$ are determined from $|F_N|$, squared, and used as coefficients in a Patterson summation, so yielding a sharpened Patterson map.

To remove the origin from a Patterson map, the quantity $|F|^2$ obtained from the intensity measurements has the quantity $\sum_{i=1}^{N} f_i^2$ subtracted from it on the absolute scale. In the case of a sharpened Patterson map $f_i = Z_i$

and

$$I_{hkl} = |F_p|^2 - \sum_{i=1}^{N} Z_i^2$$

In certain special cases the interpretation of Patterson maps may be helped by comparing the maps produced by one crystalline sample, and a similar one where an atomic species in the first sample has been isomorphously replaced. It should then be possible to identify vector peaks arising from the isomorphous replacement.

The volume which lies around the origin of a vector map for a radius of about 2·5 Å, has a special significance in that all vectors lying within it must be due to interatomic rather than intermolecular distances in the unit cell. A close examination of this region on a Patterson map often yields important information.

In a unit cell containing several molecules where the co-ordinates of the heavy atoms have already been found, it is possible to identify the positions of light atoms by using a procedure described by Beevers and Robertson (*Acta. Cryst.*, **3**, 164 (1950)). If we consider a crystal containing four molecules per unit cell, there will be four vectors between any light atom and each of the heavy atoms. If four copies of the Patterson map are overlayed (i.e. a superposition method) with their origins at the positions of each of the heavy atoms, then the point where four peaks, one from each Patterson map, coincide is the position of a light atom.

Further aids to unravelling vector maps are:

(a) To look for the vector peaks due to some known geometrical detail of the structure, for example, benzene rings.

(b) To correlate the physical properties of the crystals with structural features. For example, prominent cleavage planes in a crystal may suggest that flat molecules are lying in these planes. In addition, the positions of large molecules may be inferred from the dimensions of the unit cell, as it may be physically impossible for certain molecules to adopt particular orientations as a result of geometrical considerations.

In some instances it is not possible to determine all the co-ordinates of an atom from the vector map. For example in the space group $P2_1$ (see Harker sections, p. 129) which has a screw axis parallel to the b-axis of the unit cell, only the x and z co-ordinates of the atom can be obtained leaving the value of the y co-ordinate unknown. In this particular case the value of y may be assigned arbitrarily, as the position of the origin on the b-axis is not at any specific position. If y is assigned the value $\frac{1}{4}$, then the two heavy atoms will be at points related by a centre of symmetry i.e. $x, \frac{1}{4}, z; \bar{x}, -\frac{1}{4}, \bar{z}$. This means that even though the space group is non-centrosymmetric the heavy atoms are related by a centre, and this leads to difficulties when the Fourier map based on these heavy atom positions has to be interpreted, as it will contain both the real structure and its mirror image.

Vector-Search Methods

Increasing use[57] is being made nowadays of a combination of both Patterson and direct methods (for a description of the latter see later) in tackling difficult structures containing light atoms only. The reader's attention is also drawn to the double-Patterson function[74] which is defined in six-dimensions, and the coefficients of which are the triple-products of direct methods.

In some cases Patterson superposition techniques are in themselves sufficient to deconvolute the Patterson map and obtain one image of the structure from the N images of the map. In other cases use must be made of any additional information that is available such as chemical knowledge, where the shapes of certain groups in the structure are known, or symmetry information, which can be obtained from the space-group. Where a group of known shape exists in the molecule, a vector-search technique may be employed to look for known vectors in the Patterson map and so ultimately elucidate the rest of the structure.

R. A. Jacobson[57] describes one approach to the problem of incorporating symmetry information into the Patterson method by means of a symmetry minimum function. In addition a procedure known as vector verification is used to check all possible atomic positions for consistency with all Patterson vectors[58, 59, 60]. The Patterson function is computed using the fast Fourier program which incorporates the Cooley-Tukey algorithm[61]. Suitable peaks are then selected for the application of the superposition method, which works well as long as spurious peaks are not chosen in the early stages.

Two further methods of peak-evaluation are described which do not rely on chemical information; one involves the calculation of a 'frequency check'[62] and the other the calculation of a 'discriminator function'[63].

The correctness of fit of a structural model or fragment of such, can be measured in terms of a sum, product, or minimum function[64], i.e.

$$\sigma(\theta_1, \theta_2, \theta_3) = \sum_{ij} P(\mathbf{r}_i - \mathbf{r}_j) \quad (5.3)$$

$$P(\theta_1, \theta_2, \theta_3) = \Pi_{ij} P(\mathbf{r}_i - \mathbf{r}_j) \quad (5.4)$$

$$M(\theta_1, \theta_2, \theta_3) = \text{Min}_{ij} P(\mathbf{r}_i - \mathbf{r}_j) \quad (5.5)$$

$\mathbf{r}_i$ are the relative co-ordinates of the atoms of the known fragment. $\mathbf{r}_i - \mathbf{r}_j$ is the theoretical vector set for this fragment, where $i = 1, 2 \dots n$ and $j = 1, 2, \dots n$ measured over the vector $\mathbf{r}_i - \mathbf{r}_j$. θ_1, θ_2, and θ_3 are the Eulerian angles used to specify rotation[65, 66].

The correct interpretation of the Patterson map then resolves itself into finding a matrix $[C]$ and a vector $\mathbf{t}$ such that

$$[C]\mathbf{r}_i + \mathbf{t} = \mathbf{R}_i$$

gives the position vectors R_i of the atoms with respect to the unit cell of the crystal. The orientation of the known group is found when $[C]$ is known and the position of the group is found when t the translation vector is known. The application of equations (5.3), (5.4) or (5.5) or other more complicated functions[67, 68, 69] then tests the validity of the interpretation of the Patterson map.

M. J. Buerger[70] discusses image-seeking functions (sum, product, and minimum) and states the following morals applicable to their use:

1. If possible always start an image-seeking function by using a single point, i.e. a point at which two images are not superposed.
2. Do not 'sharpen' your Patterson function and then wonder why your use of the minimum function did not lead to a successful conclusion.

P. Tollin in a discussion of Patterson Function Interpretation Techniques[64] reviews applications of image-seeking functions in both reciprocal- and real-space, and, as mentioned above, the problem resolves itself into finding (a) the orientation of the known group and (b) the location of the known group, i.e. the matrix $[C]$ needed to calculate $r_i = [C]r_i$, and the translation vector t. The components of the vector are X_0, Y_0, Z_0, and the procedure is as follows. If r_i are the correct relative co-ordinates, T is a symmetry operator, and the molecule is at $(r_i + t)$, then there will be intermolecular vectors such as $[(r_i + t) - T(r_j + t)]$ for all i and j. The Patterson function is then examined at these locations, i.e. $P[(r_i + t) - T(r_j + t)]$ for all i and j and all values of t. The best fit of the vector set to the Patterson function as t varies is what is required.

A sum function approach can be used to find the correct *orientation* of the molecule where equation (5.3) is re-written:

$$\sigma(\theta_1, \theta_2, \theta_3) = \sum_h |F_{(h)}|^2 \left[\left(\sum_i \cos^2 \pi h r_i \right)^2 + \left(\sum_i \sin 2\pi h r_i \right)^2 \right]$$

An alternative function for determining the orientation of the molecule is the rotation function, which is particularly useful in looking at structures of great complexity. It is then used in conjunction with the sum function (which if used alone would require too much computing time).

The rotation function[65],

$$R(\theta_1, \theta_2, \theta_3) = \int_u P_2(x_2) P_1(x_1) \, dx_1$$

measures the degree of coincidence within the volume u when the Patterson function P_1 is rotated on the Patterson function P_2.

A point x_1 in P_1 is related to any other point x_2 in P_2 by the rotation matrix C, i.e.

$$x_2 = [C]x_1$$

Yet another function, the $I(\theta\phi)$ function[64], may be used in finding the orientation of a planar group. The orientation is described in terms of spherical polar angles θ and ϕ and an azimuthal angle ψ, rather than by Eulerian angles as in the other functions that have been mentioned.

Once the orientation of the group is known, i.e. the vectors r_j which define the relative positions of the n atoms of the group, it remains to find the three co-ordinates (X_0, Y_0, Z_0) which specify the vector R_0, which defines the position of one of the atoms with respect to the origin of the unit cell.

To determine the translation vector a Q function[71] can be used which may be described in terms of the Patterson function, when

$$Q(R_0) = \sum_{ij} P[(r_i + R_0) - T(r_j + R_0)] \tag{5.6}$$

or the image-seeking sum function when,

$$Q(R_0) = \sum_{j=1}^{n} S_n[T(r_j + R_0) - R_0] \tag{5.7}$$

i.e. the sum function $S_n(r) = \sum_{j=1}^{n} P(r - r_j)$

A symmetry element $T(r)$ relates an atom of the group whose true position is $r_j + R_0$ to a symmetry related atom at $T(r_j + R_0)$ (see earlier).

A similar function to equation (5.6), a translation function, has also been proposed[72].

The descriptions so far of image-seeking functions have all been ones which are used in reciprocal space.

A real space approach has been described by W. Hoppe[69, 73] where the concept of a *convolution molecule* is used to interpret Patterson functions.

The Patterson function may be described as the convolution of the electron density with itself inverted in the origin.

$$P(r) = \int \rho(r^1)\rho(r + r^1) \, dr^1$$

The Fourier transform is then the product of $F(h)$ and $F(\bar{h})$. If there are n molecules in the structure described by $\rho_i(i = 1, \ldots n)$ then the

Patterson function consists of the sum of all the convolution products,

$$\rho_i * \rho_j (i, j = 1, \dots n)$$

These products are convolution molecules, which may be of two types, mixed index ($\rho_i * \rho_j$) and equal index ($\rho_i * \rho_i$). The former represent intermolecular vectors and the latter intramolecular vectors, and a similarity may be noted here between this approach and the reciprocal space approach, where separation into orientation determination and translation occurs.

The application of the convolution molecule technique proceeds via, first, the calculation of an equally indexed convolution molecule using the relative fractional co-ordinates of the known fragment (the orientation of the fragment); and then a mixed index convolution molecule is calculated for each of the symmetry related molecules. Criteria of fit are obtained using functions that differ in some respects from those mentioned earlier.

The reader who wishes to read further about the solution of partially solved structures is referred in addition to G. N. Ramachandran, *Advanced Methods of Crystallography*, Academic Press (1964).

The Fourier Synthesis

When the point is reached in a structure analysis where it is possible to compute a Fourier synthesis and so obtain an electron density contour map of the unit cell contents, the structure may be considered virtually solved, because at this point both the magnitude and the phase of the structure factors will be known.

As with the Patterson synthesis, it is usual to calculate a Fourier synthesis only for the asymmetric part of the unit cell, which will contain all the electron density information of the crystal structure. If the heavy-atom method has been employed to obtain the structure factor phases, then the input for the Fourier synthesis may be obtained from the output of the refinement of the heavy-atom parameters. If direct methods have been employed then Fourier syntheses are used to produce E-maps (see the section on direct methods).

It is important to bear in mind the relationship between the calculation of structure factors and the calculation of a Fourier synthesis. Structure factors may be calculated for a known distribution of electron density. Fourier syntheses provide an electron density distribution from a set of structure factors. The electron density is the Fourier transform of the structure factors and vice versa.

Theory

A crystal is a periodic structure and as such may be described by means of a periodic function, the Fourier series*, as suggested by W. H. Bragg in 1915. If first, a hypothetical one-dimensional crystal is considered, the electron density distribution may be described by the following Fourier series:

$$\rho(x) = \sum_h C(h)\, e^{2\pi i h x} \tag{5.8}$$

where $\rho(x)$ is the electron density at point x, and $C(h)$ is the Fourier coefficient.

The expression for the structure factor is:

$$F(h^1) = \int \rho(x)\, e^{2\pi i h x}\, dx \tag{5.9}$$

where the integral is over the length a of the one-dimensional unit cell.

Substituting equation (5.9) in equation (5.8):

$$F(h^1) = \int \sum_h C(h)\, e^{2\pi i h x}\, e^{2\pi i h^1 x}\, dx$$

The integral is zero unless $h = -h^1$ when the exponential terms disappear and,

$$F(h) = \int C(h)\, dx$$

therefore $C(h) = F(h)/a$, i.e. the coefficients of the Fourier series are proportional to the corresponding structure factors and equation (5.8) becomes

$$\rho(x) = 1/a \sum_h F(h)\, e^{-2\pi i h x}$$

If the structure is centrosymmetric:

$$\rho(x) = 1/a \sum_h F(h) \cos 2\pi h x$$

In an analogous way it can be shown that for the three-dimensional

*For a discussion of this see, for example, reference 75.

case, the electron density at the point x, y, z is

$$\rho(x, y, z) = 1/V \sum_h \sum_k \sum_l F_{(hkl)} e^{-2\pi i(hx + ky + lz)}$$

(where V = unit cell volume) which becomes, for the centrosymmetric case:

$$\rho(x, y, z) = \frac{1}{V} \sum_h \sum_k \sum_l F_{(hkl)} \cos 2\pi(hx + ky + lz)$$

The phase angle, α, may be introduced in the non-centrosymmetric expression, when the structure amplitude $|F_{(hkl)}|$ replaces the structure factor $F_{(hkl)}$, and,

$$\rho(x, y, z) = 1/V \sum_h \sum_k \sum_l |F_{(hkl)}| \cos \{2\pi(hx + ky + lz) - \alpha_{(hkl)}\}$$

α is defined by, $\tan \alpha_{(hkl)} = B/A$ where $A = |F| \cos \alpha_{(hkl)}$, $B = |F| \sin \alpha_{(hkl)}$, and $F_{(hkl)} = A_{(hkl)} + iB_{(hkl)}$

For computational purposes, the above summations need only be carried out over half of reciprocal space as long as Friedels Law holds good, i.e. $F(hkl) = F(\bar{h}\bar{k}\bar{l})$. However, the zero order term $F_{(000)}$ is its own conjugate, and is therefore treated separately. The expression for the electron density is then[76]:

$$\rho(xyz) = 1/V[F_{(000)} + 2\sum_{h=0}^{\infty} \sum_{k=-\infty}^{\infty} \sum_{l=-\infty}^{\infty} |F_{(hkl)}| \cos \{2\pi(hx + ky + lz) - \alpha_{(hkl)}\}] \qquad (5.10)$$

The value $F_{(000)}$ is a constant equal to the number of electrons in the unit cell when absolute structure amplitudes are used in the expression for $\rho(xyz)$.

For the two-dimensional case the above expression becomes:

$$\rho(xy) = 1/A [F_{(000)} + \sum\sum |F_{(hk0)}| \cos \{2\pi(hx + ky) - \alpha_{(hk0)}\}]$$

and for the projection of the electron density on to a line, e.g. a cell edge, the expression is:

$$\rho(x) = 1/d_{100} [F_{(000)} + \sum |F_{(h00)}| \cos \{2\pi(hx) - \alpha_{(h00)}\}]$$

Ideally a Fourier summation should contain an infinite number of terms. In practice, there is a limit to the amount of data that can be observed. In addition there will be errors in the observed structure amplitudes and phases. Consequently the electron density map calculated using a Fourier synthesis will not be a perfect representation of the distribution of electron density in the crystal, and therefore the atomic positions, once obtained, are refined, usually by a least-squares procedure (see later).

For example, it may only be possible to collect intensity data in the range $\pm h$, $\pm k$, $\pm l = 10$, instead of from $+\infty$ to $-\infty$ for each index. The errors in the electron density map arising from this are known as termination-of-series errors.

When a Fourier synthesis is carried out, the points x, y, z at which the electron density is to be calculated must be specified. For example, the unit cell may be split up into one-sixtieths of each cell edge and a three-dimensional net obtained, where the electron density has been calculated at each corner of the cell sub-units. This is done in practice by computing sections of the unit cell up one particular axis, so that a series of two-dimensional nets are obtained, on which electron density contours may be drawn.

An Example of the Use of Patterson and Fourier Syntheses — structure of diethyl(salicylaldehydato)thallium(III) *J. Chem. Soc.* (A) 648 (1967).

This particular structure was solved in the non-standard space group $C\bar{1}$, i.e. a C-face centred triclinic cell having a centre of symmetry. There are four formula units, $Et_2(C_7H_5O_2)Tl$ per unit cell.

Patterson projections were calculated down each of the crystallographic axes, i.e. down the [100], [010], and [001] axes. The expected large peaks representing thallium–thallium atom vectors were

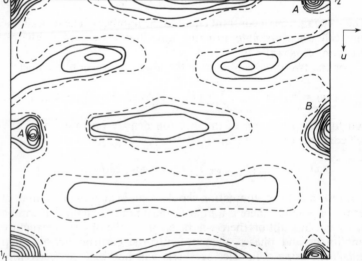

Figure 5.9. [001] *Patterson projection; the zero contour is shown by a dotted line*

only observed in the latter case which suggested that in the other two projections these peaks were concealed in the origin peak. Figure 5.9 shows the projection down the [001] direction for half the unit cell; the thallium–thallium peaks are labelled A and the large peak B arises from the C-face centring. Figure 5.10 shows the section $W = 0$ from a full three-dimensional Patterson synthesis and the heavy

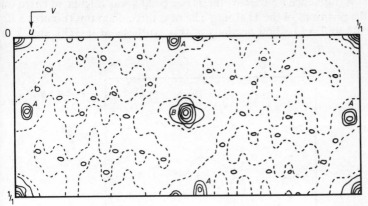

Figure 5.10. Patterson section at $W = 0$; the zero contour is shown by a dotted line

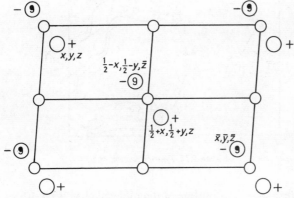

Figure 5.11. Equivalent positions for the space group $C\bar{1}$

atom peaks, labelled A, are easily seen. The co-ordinates of the Tl atom in the unit cell are easily shown to be $x = 7/30$ of the cell edge, $y = 1/60$ of the cell edge, and $z = 0$, from the information obtained from the Patterson synthesis.

Figure 5.11 shows the equivalent positions in the unit cell and, as the Tl atom z-co-ordinate is zero, it follows that four of the Tl atoms lie in the C-faces of the unit cell.

In addition to obtaining the Tl atom co-ordinates it could be seen from the Patterson synthesis that most of the vector peaks were contained in the sections from 0 to 7/30 along W. This information is useful when examining the Fourier synthesis based on the Tl atom positions for the electron density peaks of the other atoms in the molecule.

A difference Fourier synthesis (*see* p. 200) was calculated based on the positions of the Tl atoms. The next three diagrams (Figures 5.12, 5.13 and 5.14) show sections of this synthesis at 0, 3/30, and 7/30

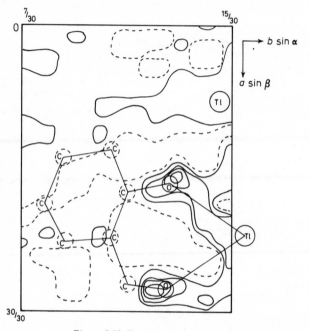

Figure 5.12. Fourier section at z = 0

along the z-axis. The outline of the molecule has been drawn in and it can be seen that as one goes from one section to the next that more atoms in the salicylaldehyde ligand become visible. The broken circles represent the positions of atoms which appear on the later sections. The unbroken circles are the atoms which have already appeared. The section at 7/30 along the z-axis shows the Tl atom (having fractional co-ordinates $x = 0.73$, $y = 0.52$, and $z = 0$) and its co-ordinated salicylaldehyde molecule, which makes an angle of $\sim 20.6°$ to the sections.

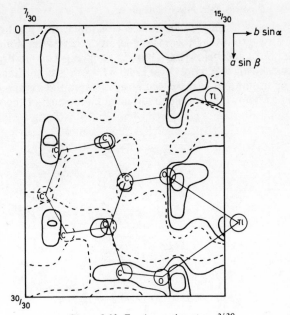

Figure 5.13. Fourier section at z = 3/30

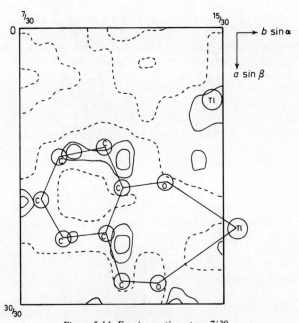

Figure 5.14. Fourier section at z = 7/30

At this point it only remained to determine the positions of the two ethyl groups in the molecule, and these were found, as expected, above and below the Tl atom, so giving a six-co-ordinate Tl atom as shown in Figure 5.15. Each Tl atom is co-ordinated to two carbon atoms and four oxygen atoms; two of the oxygen atoms belonging to its own salicylaldehyde ligand, and each of the others

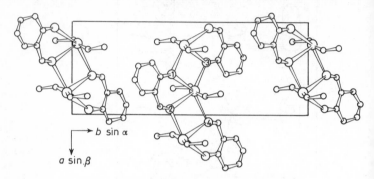

Figure 5.15. View of the unit cell down the c-axis

belonging to salicylaldehyde ligands of adjacent Tl atoms. The oxygen atoms are numbered 1, 2, 3, and 4 in the diagram.

Discussions of Fourier synthesis theory and its applications are given in several text books, e.g. *Crystal Structure Analysis*, M. J. Buerger; *X-ray Structure Determination*, G. H. Stout and L. H. Jensen; *The Determination of Crystal Structure*, H. Lipson and W. Cochran.

STRUCTURE REFINEMENT BY LEAST-SQUARES METHODS

The Structure Factor

In Cruickshank's[77] summary of the structure refinement process it can be seen that if the electron density is a superposition of j 'atomic' electron densities, then the structure factor $F_{(hkl)}$ is given by:

$$F_{(hkl)} = \sum_j f_j(hkl) \exp\left[2\pi i(hx_j + ky_j + lz_j)\right]$$

where the summation is over all the atoms in the unit cell and

$$f_j(hkl) = V \iiint_{-\infty}^{+\infty} \rho_j(uvw) \exp\left[2\pi i(hu + kv + lw)\right] dudvdw$$

is the scattering factor of atom j whose electron density is $\rho_j(uvw)$; (uvw) being co-ordinates referred to (x_j, y_j, z_j) as origin.

If the atomic densities are centrosymmetric with respect to their origins, the f_j are real and the real and complex components of $F(hkl)$ are:

$$A(hk) = \sum_j f_j(hkl) \cos \left[2\pi(hx_j + ky_j + lz_j)\right] \qquad (5.11a)$$

$$B(hk) = \sum_j f_j(hkl) \sin \left[2\pi(hx_j + ky_j + lz_j)\right] \qquad (5.11b)$$

In space groups of higher symmetry than $P1$ the summation over all atoms j is usually split into a summation over symmetry related atoms followed by a summation over the members of the asymmetric unit. Space groups having a centre of symmetry at the origin have a value of zero for $B(hkl)$.

International Tables, Vol. I, gives simplified forms of the trigonometric summations over symmetry related atoms in equations (5.11a) and (5.11b). These expressions are valid only for atoms possessing at least the symmetry elements of the point group, but not for atoms subject to general anisotropic vibrations. Modern computer programs do not make use of these expressions, and symmetry may be dealt with in terms of a 3×3 rotation matrix R and a translation vector t.

Least-Squares Refinement

The term *structure refinement* refers to an iterative process where a trial set of parameters are modified by a least-squares procedure, to give the best fit between a set of observed and a set of calculated data.

The parameters that are adjusted include the atomic co-ordinates, the atomic vibration parameters, and a scale factor which is used to put the observed structure factors on an absolute scale.

The observed data are the structure factors obtained from the intensity data, whose phases have been determined by direct or indirect methods.

The Calculated Structure Factor, F_c

The calculated data are the structure factors obtained from the expression

$$F_{(hkl)} = \sum_j f_j(hkl) \exp \left[2\pi i(hx_j + ky_j + lz_j)\right]$$

When a centre of symmetry is present in the crystal space group, e.g. space group No. 2, $P\bar{1}$, the above expression reduces to:

$$F_{(hkl)} = \sum_j f_j(hkl) \cos 2\pi(hx_j + ky_j + lz_j)$$

This expression is now real, not complex, and wherever possible the origin of the unit cell is chosen at a centre of symmetry so that real structure factors are obtained.

The Observed Structure Factor, F_o

The magnitude of the observed structure factor comes from the intensity measurements after they have been corrected for Lorentz-polarisation effects, possibly absorption and extinction effects, and for photographic data, after a Philips spot-shape correction has been applied. The phases of the observed structure factors are obtained mathematically by direct-methods, or by Patterson and Fourier syntheses if the heavy-atom method has been employed.

The Atomic Scattering Factor, f_n

When the theory of diffraction by a crystal lattice is considered the scattering units are usually treated as single electrons so that their linear dimensions may be ignored. In this way the scattering of X-rays is made independent of the scattering angle, and an atom which is considered to have all its electrons concentrated at one point will display no destructive interference in the rays scattered from different electrons. The scattering power, or form factor, f, of such an atom would show no variation of f with $\sin\theta/\lambda$. As f is expressed in terms of the scattering power of a single electron, its value would remain constant at Z, the atomic number of the atom.

When a real atom is considered the phase differences in the rays scattered from various parts of the electron cloud are small at small diffraction angles, and the total scattering amplitude is the sum of the amplitudes of the rays scattered by the individual electrons, i.e.

$$f = Z \quad \text{when} \quad \sin\theta/\lambda = 0$$

At larger diffraction angles the destructive interference of rays scattered by different electrons in the atom increases and the magnitude of f decreases with increase of $\sin\theta/\lambda$. Values of f for all species of atoms as a function of $\sin \theta/\lambda$ are given in standard texts (see *International Tables for X-ray Crystallography*, Vol. III p. 201).

Anomalous Dispersion

For both centrosymmetric and non-centrosymmetric crystals the intensities of diffracted rays from planes (hkl) and $(\bar{h}\bar{k}\bar{l})$ are identical (i.e. Friedel's law) unless one of the atoms causing the diffraction has an absorption edge just on the long wavelength side of the radiation used for diffraction. When this occurs the scattered wave exhibits an anomalous phase shift. The method of allowing for anomalous dispersion is to apply two additional factors to the scattering factor curve, i.e. $f = f_0 + f^1 + if^{11}$. To apply f^1, which is referred to as the real part of the anomalous dispersion, a certain number of electrons are usually subtracted from the scattering factor of the particular atom for the whole $\sin \theta/\lambda$ range. To apply f^{11}, the imaginary component, is more difficult and in fact it is usually much smaller in its effect on f_0 than is f^1. Values of f^1 and f^{11} for different elements and radiations are given in *International Tables for X-ray Crystallography*, Vol. III, p. 213–216.

Bijfoet[78-82] has made use of anomalous scattering to determine absolute configurations, and the allocation of phases to structure factors can be made using anomalous dispersion methods[83, 84, 85]. (Also see 'other approaches to the phase problem' later.)

The Temperature Factor (Debye–Waller factor)

The description given above of atomic scattering factors makes no allowance for the thermal vibrations of the atom. These will effectively increase the volume occupied by the electrons producing the scattering of the X-rays. The result is a more rapid fall-off of f, with $\sin \theta/\lambda$, than would be obtained for an atom at rest. To allow for this effect the scattering factor f_0 is multiplied by a temperature factor, so that[86]:

$$f = f_0 \cdot e^{-(B\sin^2 \theta)/\lambda^2}$$

B is the temperature coefficient which is related to the mean square displacement of the atoms from their mean position by the expression $B = 8\pi^2 \overline{u^2}$.

If the structure is known and values of observed and calculated intensities are available, then B can be obtained from the expression:

$$I_0 = I_c \exp\left(-2B\sin^2 \theta/\lambda^2\right)$$

A graph of $\log I_0/I_c$ against $\sin^2 \theta/\lambda^2$ has a slope $2B$ [86].

Usually, however, a value of B is required before the structure is known. A scale factor is also required during a structure analysis to put the observed structure factors on an absolute scale. A method is described below of obtaining both the scale factor and the overall temperature coefficient B.

The Temperature Coefficient B and the Scale Factor

Wilson[87] has shown that the intensity of an X-ray reflection may be described by the expression:

$$I_{(hkl)} = \sum_a f_a^2 + \sum_a \sum_b f_a f_b \exp \{2\pi i[h(x_a - x_b) + k(y_a - y_b) + l(z_a - z_b)]\}$$

where $a \neq b$.

If $I_{(hkl)}$ is averaged for all reflections in a given range of $\sin^2 \theta/\lambda^2$ and it is assumed that $\lambda/\sin \theta$ is small compared to the interatomic distances, then the exponential terms in the above expression which will have both positive and negative values will, on averaging over hkl, be practically zero. Therefore:

$$I_{(hkl)} = \sum_a \overline{f_a^2} \tag{5.12}$$

where $\overline{f}$ are the atomic structure factors for the centre of the $\sin^2 \theta/\lambda^2$ range. I_0 will differ from I_c by two factors, a scale factor which is independent of θ, and a temperature factor. The ratio $\overline{I}_0/\overline{I}_c$ for the group of reflections will be the product of these factors, c. By dividing the reflections into groups of suitable ranges of $\sin^2 \theta/\lambda^2$ and obtaining $\overline{f}^2/\overline{I}_0$ for each group, the factor $1/c$ is found as a function of $\sin \theta/\lambda$. If $\log 1/c$ is plotted against $\sin^2 \theta$ a straight line is obtained if the normal form of the temperature factors is correct, i.e. $\exp(-B \sin^2 \theta/\lambda^2)$. The axial intercept of the curve gives a value $1/c$ and the slope $= B$.

The values of the scale factor and temperature factor may be calculated at the beginning of a structure analysis in the following way. The values of $|F_0|^2$ are usually on an arbitrary scale so that:

$$|\overline{F}_0|^2 = K|\overline{F}|^2$$

Now

$$|\overline{F}|^2 = \sum_j f_j^2$$

and

$$f^2 = f_0^2 \cdot e^{-(2B \sin^2 \theta)/\lambda^2}$$

Therefore substituting and taking logs,

$$\ln\left(\frac{|F_0|^2}{\sum_j f_{0j}^2}\right) = \ln K - \frac{2B}{\lambda^2}\sin^2\theta$$

This equation has the form $y = mx + c$, i.e. a straight line, and if values of $\sin^2\theta$ are plotted against $\ln(|\bar{F}_0|^2/\Sigma f_{0j}^2)$ the value of $\ln K$ is obtained from the axial intercept and the gradient $= -2B/\lambda^2$. The above expression can be rearranged to give a value of B.

$$-\left[\ln\left(\frac{|F_0|^2}{\sum_j f_{0j}^2}\right) - \ln K\right]\cdot\frac{\lambda^2}{2\sin^2\theta} = B$$

The usual procedure that is followed when preparing the data for a graph is to split the data into a number of zones of the reciprocal lattice each having a characteristic range of $\sin^2\theta$, e.g. from 0 to 1·0 in steps of 0·2.

It is assumed that in each range (e.g. 0·0–0·2) sufficient averaging occurs for the assumptions made in obtaining Eq. (5.12) to hold good. The average value of $|F_0|^2$ is then divided by $\sum_j f_{0j}^2$ for each range of $\sin^2\theta$.

Atomic Vibration Parameters

The parameter B that was described above is the overall temperature coefficient for the structure. It is related to the magnitude of vibration of the molecule, and its use implies that all the atoms in the molecule will have the same amplitude of vibration.

When a structure is being refined better results will obviously be obtained if individual atomic vibration parameters can be used. In the initial stages of refinement each atom is allotted an isotropic vibration parameter, which assumes that the vibrations are spherically symmetrical. In the later stages of refinement anisotropic vibration parameters, six in all, are used. In this way the vibrations of the atom are described by an ellipsoid of vibrations.

In Cruickshank's[77] summary of the refinement process it can be seen that the scattering factor, f_t, of an atom in thermal vibration may be described as the product of f for the atom at rest multiplied by q, the transform of the 'smearing' function t, i.e.

$$f_t(hkl) = f(hkl)\cdot q(hkl)$$

In the isotropic case the smearing function $t(x)$ is a Gaussian and

$$t(x) = (2\pi^2 U)^{-3/2}\exp(-x^2/2U)$$

The only parameter in the expression is $U = \overline{u^2}$ the mean square amplitude of vibration in any direction.

The transform of $t(x)$ for isotropic vibration is $q(s) = \exp[-B(\sin\theta/\lambda)^2]$ where $B = 8\pi^2 U = 8\pi^2\overline{u^2}$ the Debye factor.

Anisotropic atomic vibrations can be described by a symmetrical tensor U, having six independent components such that the mean square amplitude of vibration in the direction of a unit vector $l(= l_1, l_2, l_3)$ is:

$$\overline{u^2} = \sum_{i=1}^{3}\sum_{j=1}^{3} U_{ij}l_il_j$$

or

$$\overline{u^2} = U_{11}l_1^2 + U_{22}l_2^2 + U_{33}l_3^2 + 2U_{23}l_2l_3 + 2U_{31}l_3l_1 + 2U_{12}l_1l_2$$

The factor 2 arises as $U_{23} = U_{32}$ etc. The terms U and l are described with respect to the reciprocal axes a^*, b^*, c^*, so that the component of U in the [100] direction parallel to a^* is $\overline{u^2} = U_{11}$.

In the anisotropic case the transform of the smearing function is:

$$q(S) = \exp[-2\pi^2(\Sigma\Sigma U_{ij}S_iS_j)]$$

where $S = (S_1, S_2, S_3)$ and is the reciprocal vector.

At a reciprocal lattice point $S = (ha^* + kb^* + lc^*)$

$$q(hkl) = \exp[-2\pi^2(U_{11}h^2a^{*2} + U_{22}k^2b^{*2} + U_{33}l^2c^{*2} +$$
$$2U_{23}klb^*c^* + 2U_{31}lhc^*a^* + 2U_{12}hka^*b^*)]$$

The units of the U_{ij} are Å^2 and the six B_{ij} describe an ellipsoid of vibration.

An alternative way of looking at the anisotropic vibration parameters[88] is to rearrange the expression for the isotropic temperature factor, e.g.

$$\exp\left[-B\left(\frac{\sin\theta}{\lambda}\right)^2\right] = \exp\left[-\frac{B}{4}\left(\frac{2\sin\theta}{\lambda}\right)^2\right]$$

$$= \exp\left[-\frac{B}{4}\left(\frac{1}{d}\right)^2\right]$$

where d = the interplanar spacing. Therefore, $1/d_{(hkl)}$ is the length of a reciprocal lattice vector from the origin to the point hkl.

Now

$$1/d_{(hkl)} = (h^2a^{*2} + k^2b^{*2} + l^2c^{*2} + 2hka^*b^*\cos\gamma^* + 2hla^*c^*\cos\beta^* + 2klb^*c^*\cos\alpha^*)^{\frac{1}{2}}$$

The temperature factor which will have a component for each of the above terms may then be written:

$$\exp[-\tfrac{1}{4}(B_{11}\ldots + B_{22}\ldots + B_{33}\ldots + 2B_{12}\ldots + 2B_{13}\ldots + 2B_{23}\ldots)]$$

The B_{ij} are thermal parameters having the same units as the temperature coefficient B.

The Process of Refinement

A refinement program based on the least-squares procedure calculates structure factors and accumulates least-square totals which are later solved for the parameter changes. The parameters are treated as functions of $|F_c|$ and the least-squares procedure minimises the function $\Sigma w(|F_0| - |F_c|)^2$ where w is a weight allocated to each structure factor (see later). $(|F_0| - |F_c|)$ is represented by Δ and the function minimised is then $w\Delta^2$.

The parameters that are usually refined are:

1. The $|F_0|$ scale factor.
2. The atomic co-ordinates.
3. Either the individual atomic isotropic vibration parameters, or the six components of the individual atomic anisotropic vibration tensors, U_{ij}.

Small variations are made to these parameters to produce the best agreement between observed and calculated structure factors. The least-squares method, originally described by Legendre[89] proceeds by making the sum of the squares of the errors in F_0 a minimum. A cyclic process is carried out and after each cycle an improved value for each parameter is obtained. The procedure is repeated until no further improvement takes place, as shown by an agreement index $R = \Sigma(|F_0| - |F_c|)/\Sigma|F_0|$. This may be expressed as a percentage or a decimal fraction, e.g. 10% or 0.10.

Cruickshank[90] has shown that to obtain bond lengths to within an error of 0.01 Å, the final agreement index must have a value of about 1%. The usual value of R obtained by film methods varies from about 7 to 10% for a well refined structure (a better agreement index is obtained when diffractometer data are used). The advisability of continuing the refinement can also be gauged from a comparison of the shifts in the parameters produced by the program compared with the corresponding estimated standard deviations.

In the summary by Cruickshank[77] the normal equations which must be solved for the small changes ε_j to any parameter ρ_j are shown to be:

$$\sum_{i=1}^{n} \varepsilon_j \left\{ \sum_{hkl} w \frac{\partial|F_c|}{\partial \rho_j} \cdot \frac{\partial|F_c|}{\partial \rho_i} \right\} = \sum_{hkl} w\Delta \frac{\partial|F_c|}{\partial \rho_j}, (j = 1, \ldots n)$$

This may be expressed in matrix form:

$$\sum_i a_{ij}\varepsilon_i = b_j, \quad j = 1, \dots n$$

where

$$a_{ij} = \sum_{hkl} w \, \frac{\partial |F_c|}{\partial \rho_i} \cdot \frac{\partial |F_c|}{\partial \rho_j}$$

$$b_j = \sum_{hkl} w\Delta \, \frac{\partial |F_c|}{\partial \rho_j}$$

Although it is preferable, it is not necessary to calculate a full least-squares matrix to obtain the values of ε_i. Such a calculation would need the use of a large computer store. A block-diagonal approximation may be used and this has been shown to be preferable to a simple diagonal approximation for the following reasons[91].

It is assumed when using a diagonal approximation that as the a_{ii} elements in an a_{ij} matrix will all be squares and therefore positive, the off-diagonal elements which will be products and therefore may be positive or negative, may be ignored, since an accumulation of diagonal terms contributing to the a_{ii} elements will be much larger than the accumulation of off-diagonal terms contributing to the a_{ij} elements.

However, this argument overlooks the fact that there may be correlation between various parameters leading to large positive or negative values of the a_{ij} elements of the matrix. The diagonal approximation is known to give poor results for the scale factor and the overall vibration parameter $\bar{U}$, which are always coupled when refined. Similarly poor results are obtained for atomic co-ordinates in non-orthoganol systems and for anisotropic vibrations.

The block-diagonal approximation retains the off-diagonal elements where correlations exist, e.g. the $|F_0|$ scale and the average vibration, $\bar{U}$, are treated in a 2×2 matrix. The atomic co-ordinates of each atom are treated in a 3×3 matrix, and the anisotropic vibration parameters for each atom are treated in a 6×6 matrix. In this way important off-diagonal elements are retained while a full matrix least-squares calculation is avoided.

Weighting Functions

Each observation has an associated weight which is used in the function minimised by the least-squares procedure, $\Sigma w\Delta^2$. This weight should reflect the reliability of the particular measurement, and $w = 1/\sigma^2$, where σ^2 is the variance of the measurement due o random experimental errors and σ is the standard deviation.

In the initial stages of refinement a weight of unity may be used or all reflections. In the later stages of refinement it is usually necessary to introduce some form of weighting scheme which is usually dependent on F_0.

The least-squares formula[77] for standard deviations (which allows for all random experimental errors, such systematic experimental errors that cannot be paralleled in the calculated model and such defects in the model that are not paralleled in the experimental data) is shown in equation (5.13), where the weights reflect the trend in $|\Delta F|$.

$$\sigma^2(x) = (a^{-1})_{ii}(\Sigma w \Delta^2)/(m-n)^* \qquad (5.13)$$

where $(a^{-1})_{ii}$ is the appropriate element of the matrix inverse to the normal equation matrix; m is the number of observations; and n is the number of unknowns.

This is invalid unless the weights are correctly chosen. It is necessary that the $w\Delta^2$ values should be constant when analysed in a systematic manner. Most refinement programs include the facility for obtaining values of average $w\Delta^2$ for groups of F_0 arranged in increasing order of magnitude. From a study of these values a suitable weighting scheme may be chosen so that the average value of $w\Delta^2$ is a constant for each group of F_0, e.g. if F_0 magnitudes for a structure vary from 5 to 100, then groups of F_0 could be chosen as follows:

0–10, 10–20, 20–30, 30–40, 40–50, 50–60, 60–70, 70–80, 80–90, 90–100

The groups should be chosen so that a sufficient number of reflections are included in each group to provide a reasonable average, e.g. 30–60 reflections. A graph can then be plotted of mean Δ^2 against mean F_0 and a suitable weighting scheme chosen to make $w\Delta^2$ constant for each group of F_0.

The weights used with counter data arise from counting statistics. The standard deviation of a count σ, is given by $\sigma = N^{\frac{1}{2}}$, where N is the number of counts. Usually the intensity of a reflection is obtained by subtracting the background count from the peak count[92]. In this case, assuming equal counting times the standard deviation is given by:

$\sigma = (\sigma_p^2 + \sigma_b^2)^{\frac{1}{2}}$ where σ_p = peak-count standard deviation
and σ_b = background count standard deviation.

Scale Factor

By definition $F_{(000)}$ is the number of electrons in the unit cell and the scale of F_0 has to be adjusted so that $F_{0(000)} = F_{(000)}$. In any one

* For absolute weights (reflecting the precision of $|F_0|$), $\sigma^2(x) = (a^{-1})_{ii}$.

cycle of the least-squares procedure the $|F_0|$ data may not be interfered with. Instead an inverse scale factor is applied to the $|F_c|$ data during the calculation, and at the end of the calculation the $|F_0|$ data are put on the desired scale, while the $|F_c|$ data are freed of the inverse scale factor.

An Example of a Structure Refinement

Diethyl(salicylaldehydato)thallium(III)[93] is an inorganic complex, the asymmetric unit of which contains one thallium atom (the heavy atom), two oxygen atoms, and eleven carbon atoms. The structure was solved using Patterson and Fourier methods. A three-dimensional Patterson synthesis allowed the position of the Tl atom to be determined unequivocally in the centrosymmetric unit cell. The fractional co-ordinates, and cell parameters in Å and degrees were:

x/a	y/b	z/c	a	b	c	α	β	γ
0·233	0·017	0·000	8·00	20·71	7·71	100·7	101·6	88·4

At this stage use was first made of a block-diagonal least-squares refinement program. After three cycles the Tl co-ordinates changed to the following values:

x/a	y/b	z/c
0·2377	0·0164	0·0006

The isotropic vibration parameter had a value of $0·04 \text{ Å}^2$, and the agreement index was $17·7\%$. The relatively low agreement index was a measure of the influence the Tl atom had in determining the phases of the structure factors.

The positions of the other atoms were obtained from a difference Fourier based on the calculated and observed structure factors obtained from the refinement of the Tl atom. In this way a three-dimensional electron density map was obtained with the contribution due to the Tl atom removed.

Two cycles of refinement were then carried out, varying the three co-ordinates of each atom together with the individual isotropic vibration parameters and an overall scale factor. The agreement index R fell to $14·4\%$. At this point anisotropic vibration parameters were introduced for the Tl atom and two more cycles reduced R to $10·5\%$. Initially the carbon and oxygen atoms had been allocated isotropic vibration parameters, U_{ISO}, of $0·06 \text{ Å}^2$ and $0·05 \text{ Å}^2$ respectively. An examination of the results so far showed that the U_{ISO} values of all the C and O atoms lay between 0·06 and $0·09 \text{ Å}^2$, except for one C atom the value of which had risen to $0·195 \text{ Å}^2$, suggesting that its positional co-ordinates had been incorrectly determined. A difference Fourier was again calculated and the correct co-ordinates for this atom were obtained. No other unexpected peaks showed on the electron density map, which suggested the structure was substantially correct. Two more refinement cycles reduced R to 9·54%, and the co-ordinate shifts for all the atoms were less than half the corresponding standard deviations. The structure was therefore considered to be refined. (Hydrogen atoms were omitted from the refinement process.)

Throughout the procedure 1453 independent structure factors obtained from corrected intensity data were used. In the final stages of refinement the following parameters were being varied:

$$\begin{aligned}
\text{Atomic co-ordinates } 14 \times 3 &= 42 \\
\text{Vibration parameters } (1 \times 6) + (13 \times 1) &= 19 \\
\text{Scale Factors } \quad 1 &= \underline{1} \\
\text{Total number of variable parameters} &\quad \underline{\underline{62}}
\end{aligned}$$

Tables 5.1 and 5.2 show the final results obtained for the structure.

Table 5.1   ATOMIC FRACTIONAL CO-ORDINATES
(Estimated standard deviations are shown in brackets)

Atom	x/a	y/b	z/c
T1	0·7381(02)	0·5160(01)	0·0004(02)
O_1	0·5577(28)	0·4229(11)	0·0083(29)
O_2	0·9308(28)	0·4135(11)	0·0365(30)
C_1	0·9176(49)	0·3655(19)	0·1064(52)
C_2	0·5865(39)	0·2710(15)	0·0863(41)
C_3	0·4369(44)	0·3395(17)	0·1192(47)
C_4	0·4691(52)	0·2848(20)	0·2187(55)
C_5	0·6236(51)	0·2595(20)	0·2726(53)
C_6	0·7631(45)	0·2909(18)	0·2383(51)
C_7	0·7421(45)	0·3445(17)	0·1442(47)
C_8	0·7954(60)	0·6244(23)	0·3511(62)
C_9	0·8085(50)	0·5479(19)	0·2980(52)
C_{10}	0·6824(48)	0·4983(18)	0·2856(50)
C_{11}	0·6895(55)	0·4263(21)	0·3790(56)

DIRECT METHODS

The main problem facing anyone attempting to solve a structure by X-ray diffraction methods is the correct allocation of phases to the structure factors. This is the well-known phase problem. The measurable quantities of an X-ray diffraction pattern are the intensities $I_{(hkl)}$ of the diffracted rays which contain two pieces of information, the magnitude of the structure factor $F_{(hkl)}$ which is readily available as $I \propto F^2$, and the phase associated with the structure factor which is not available to the observer. In order to obtain an electron density map of the unit cell by Fourier synthesis both the phase and the magnitude of the structure factors are needed.

Two broad classes of approach to the problem exist, namely direct and indirect methods. Indirect methods are more subjective in that they depend upon the interpretation of the data by the investigator. For example, in the heavy-atom method phases are initially

Table 5.2 ATOMIC VIBRATION PARAMETERS IN $Å^2$
(Estimated standard deviations are shown in brackets)

Atom	U_{ISO} & U_{11}	U_{22}	U_{33}	$2U_{23}$	$2U_{31}$	$2U_{12}$
T1	0·0429(06)	0·0526(06)	0·0452(06)	0·0297(10)	0·0272(11)	0·0046(10)
O_1	0·056(5)					
O_2	0·056(6)					
C_1	0·069(9)					
C_2	0·050(7)					
C_3	0·061(8)					
C_4	0·075(10)					
C_5	0·073(10)					
C_6	0·066(9)					
C_7	0·061(8)					
C_8	0·088(12)					
C_9	0·071(10)					
C_{10}	0·062(9)					
C_{11}	0·078(11)					

allocated to structure factors based on the assumption that the contribution due to the heavy atom governs the phase in each case. The heavy atom cell co-ordinates are obtained from the interpretation of a Patterson vector map of the unit cell contents. Direct methods are more objective in that they depend upon the application of mathematical relationships to determine the phases of the structure factors. At the outset direct methods were applied to centrosymmetric structures where the allocation of phases is easier than in the general case, but an increasing number of non-centrosymmetric crystal structures are now being solved by these methods.

Structure Factors

A diffracted ray in a crystal can be represented by a quantity $F_{(hkl)}$, which consists of a magnitude which is the amplitude of the scattered wave, and a phase which governs the direction of the scattered wave. In the exponential form the complex quantity $F_{(hkl)}$ is given by:

$$F_{(hkl)} = \sum_j f_j \, e^{i\phi_j} \tag{5.14}$$

where the summation is over the J atoms in the unit cell, f_j = the form factor for each j atom, which represents the scattering power of the atom, and ϕ_j = the phase angle. For a discussion of the term $e^{i\phi}$ see M. J. Buerger, *Crystal Structure Analysis*. ϕ_j the phase angle can be expressed in terms of the fractional co-ordinates x_j, y_j, z_j of the atoms in the unit cell, so that:

$$\phi_j = 2\pi(hx_j + ky_j + lz_j)$$

and

$$F_{(hkl)} = \sum_j f_j \, e^{2\pi i(hx_j + ky_j + lz_j)} \tag{5.15}$$

$F_{(hkl)}$ can also be represented in terms of its real and imaginary components when[94],

$$F_{(hkl)} = F_{(hkl)} \cos\phi + i \, | F_{(hkl)} | \sin\phi \tag{5.16}$$

This is the same as saying:

$$F_{(hkl)} = (\sum_j f_j \cos\phi_j) + i(\sum_j f_j \sin\phi_j) \tag{5.17}$$

Substituting for ϕ,

$$F_{(hkl)} = [\sum_j f_j \cos 2\pi(hx_j + ky_j + lz_j)] + i[\sum_j f_j \sin 2\pi(hx_j + ky_j + lz_j)] \tag{5.18}$$

If

$$A = \sum_{j=1}^{N} f_j \cos 2\pi(hx_j + ky_j + lz_j)$$

and

$$B = \sum_{j=1}^{N} f_j \sin 2\pi(hx_j + ky_j + lz_j)$$

then

$$F_{(hkl)} = A + iB \tag{5.19}$$

If this is written $| F_{(hkl)} | = (A^2 + B^2)^{\frac{1}{2}}$ then the term $| F_{(hkl)} |$ is the structure amplitude, the magnitude of the structure factor.

Equations (5.14)–(5.19) are different ways in which the structure factor can be represented. For a centrosymmetric space group the aggregate of the sine terms is zero, and the structure factor becomes:

$$F_{(hkl)} = \sum_j f_j \cos 2\pi(hx_j + ky_j + lz_j)$$

In vector notation

$$F_{\boldsymbol{h}} = \sum_j f_j \cos 2\pi \boldsymbol{h} \cdot \boldsymbol{r}_j$$

where $\boldsymbol{r}_j$ represents the position of the jth atom in the unit cell, and $\boldsymbol{h}$ is the position in reciprocal space of the point hkl.

In the centrosymmetric case the phase can only be π or 0, i.e. the structure factor is $-F$ or $+F$. For this reason the application of direct methods to the phase problem is referred to as sign determination.

Direct methods make use of either the unitary structure factor ($U_{\boldsymbol{h}}$) or the normalised structure factor ($E_{\boldsymbol{h}}$), which are related to $F_{\boldsymbol{h}}$ in the following way:

$$U_{\boldsymbol{h}} = F_{\boldsymbol{h}}/\sum_j f_j \qquad (5.20)$$

$\sum_j f_j$ is the maximum value $F_{\boldsymbol{h}}$ can be and therefore $U_{\boldsymbol{h}}$ is given as a fraction of the maximum $F_{\boldsymbol{h}}$.

$f_j/\Sigma_j f_j$ can be equal to n_j, when $U_{\boldsymbol{h}} = \Sigma_j n_j \cos 2\pi \boldsymbol{h} \cdot \boldsymbol{r}_j$. n_j is then for $U_{\boldsymbol{h}}$ what f_j is for $F_{\boldsymbol{h}}$.

$$E_{\boldsymbol{h}} = U_{\boldsymbol{h}}/\langle U^2 \rangle^{\frac{1}{2}} \qquad (5.21)$$

A necessary preliminary to any direct method approach is the calculation of a set of unitary or normalised structure factors from the observed corrected intensity data. Once this has been done, an attempt is then made to allocate phases through the application of various mathematical relationships as described below.

Harker–Kasper Inequalities

Harker and Kasper[95] made use of Cauchy's inequality

$$\left| \sum_j a_j b_j \right|^2 \leqslant \left(\sum_j |a_j|^2 \right) \left(\sum_j |b_j|^2 \right) \qquad (5.22)$$

to show that inequalities exist between certain structure factors in centrosymmetric crystals. For example, the application of Eq. (5.22) to the centrosymmetric unitary structure factor equation produces the relationship

$$U_{\boldsymbol{h}}^2 \leqslant \tfrac{1}{2}(1 + U_{2\boldsymbol{h}})$$

Bearing in mind that the sign of $U_{2\boldsymbol{h}}$ may be $+$ or $-$, it is obvious that if $U_{\boldsymbol{h}}$ and $U_{2\boldsymbol{h}}$ are large, then $U_{2\boldsymbol{h}}$ must be positive. Two further nequalities may be derived by adding and subtracting two unitary structure factors and applying Cauchy's inequality:

$$(U_h + U_{h'})^2 \leqslant (1 + U_{h+h'})(1 + U_{h-h'}) \qquad (5.23)$$

$$(U_h - U_{h'})^2 \leqslant (1 - U_{h+h'})(1 - U_{h-h'}) \qquad (5.24)$$

Woolfson[96] shows how these may be combined and used to obtain the sign of one of U_h, $U_{h'}$, $U_{h+h'}$, when the signs of the other two are known, by using the expression

$$|U_h| + |U_{h'}|)^2 \leqslant \{1 + s(h)s(h')s(h+h') \,|\, U_{h+h'}|\}$$

$$\{1 + s(h)s(h')s(h-h') \,|\, U_{h-h'}|\}$$

to show either or both of the relationships

$$s(h)s(h')s(h+h') = +1$$

and

$$s(h)s(h')s(h-h') = +1 \quad \text{are true.}$$

$s(h)$, $s(h')$, $s(h+h')$ represent the sign of U_h, $U_{h'}$, and $U_{h+h'}$ respectively. More detailed descriptions of Harker–Kasper inequalities can be found in standard texts[96, 97].

The Multiplicity of Origins

The origin that is chosen in a unit cell belonging to the space group $P\bar{1}$, to which the co-ordinates of the structure factor equations are referred, can be any one of eight positions which correspond to the eight unique centres of symmetry in the cell and which have as co-ordinates, $0,0,0$; $\frac{1}{2},\frac{1}{2},\frac{1}{2}$; $\frac{1}{2},\frac{1}{2},0$; $\frac{1}{2},0,0$; $\frac{1}{2},0,\frac{1}{2}$; $0,\frac{1}{2},\frac{1}{2}$; $0,\frac{1}{2},0$; $0,0,\frac{1}{2}$.

If a structure factor is positive for the origin at $0,0,0$, its sign for other origins is dependent upon the parity group to which the indices hkl belong. Where h, k and l are all even it will be positive for every origin, and these structure factors are known as structure invariants. For other parities, i.e. every combination of h, k and l is odd or even, the sign of the structure factors depends upon the origin chosen. However, the product of two structure factors from the same parity group is also a structure invariant.

One outcome of the study of structure invariants is that signs (i.e. phases) may be allotted to three arbitrarily chosen non-linear structure factors as long as their hkl parity group is not even, even, even, and as long as each belongs to a different parity group. Once three signs have been allotted, the origin of the unit cell has been fixed and all other structure factor signs must be determined with respect to this origin.

Woolfson[98] gives an illustration of the procedure followed in using Harker–Kasper inequalities and structure invariants to solve the $hk0$ projection of tetraethyl diphosphine disulphide.

Sign Relationships

Sayre[99] describes a method of phase determination using sign relationships which arise from an examination of the similarity between the electron density function for a crystal composed of like-atoms and the square of this electron density function. The procedure is applicable to equal resolved atoms but holds approximately for ordinary organic crystals which contain only C, N, O and H atoms. He shows that the structure factors of such a crystal must satisfy the relationship:

$$\sum_p \sum_q \sum_r F_{(pqr)} F_{(h-p,\, k-q,\, l-r)} = V S_{(hkl)} F_{(hkl)}$$

for all values of hkl. $S_{(hkl)}$ is a function to allow for the change in atomic shape which occurs when the electron density is squared. V is the volume of the unit cell in Å^3.

It follows that the allocation of signs to structure factors of crystals of the above kind can only be correct if they satisfy the above expression. In vector form the expression becomes:

$$F_h = \frac{1}{S_h V} \sum F_{h'} F_{h-h'}$$

Sayre applies this relationship first to a hypothetical one-dimensional example, and also to the determination of the crystal structure of hydroxyproline.

Sayre's relationship and Harker–Kasper inequalities can both be used when the structure factors are large enough, to show that

$$s(h)s(h')s(h+h') = +1$$

when they are not large enough the equals sign can be replaced by probability sign, i.e.

$$s(h)s(h')s(h+h') \approx +1 \tag{5.2}$$

Much work[100] has been directed at obtaining a quantitative estimate of the meaning of the probability sign. The result obtained by Cochran and Woolfson[101] gives satisfactory results in most cases

$P_+(h, h')$ is the probability of the sign relationship being true and is given by:

$$P_+(h, h') = \tfrac{1}{2} + \tfrac{1}{2} \tanh \{(\varepsilon_3/\varepsilon^3) \,|\, U_h U_{h'} U_{h+h'} \,|\}$$

where

$$\varepsilon_3 \sum_{j=1}^{N} n_j^3, \quad \varepsilon = \sum_{j=1}^{N} n_j^2$$

For equal atoms this becomes:

$$P_+(h, h') = \tfrac{1}{2} + \tfrac{1}{2} \tanh (N \,|\, U_h U_{h'} U_{h+h'} \,|)$$

where N = the number of point atoms.

In certain circumstances there are often several pairs of known signs of type h' and $h+h'$ for a particular value of h. For the space group $P\bar{1}$ Woolfson[102] shows how the following probability relationship may be derived for these cases:

$$s(h) \approx s(\textstyle\sum_{h'} U_{h'} U_{h+h'}) \tag{5.26}$$

$P_+(h)$, the probability that U_h is positive, is then given by:

$$P_+(h) = \tfrac{1}{2} + \tfrac{1}{2} \tanh \{(\varepsilon_3/\varepsilon^3) \,|\, U_h \,|\, \textstyle\sum_{h'} U_{h'} U_{h+h'}\}$$

This expression assumes that the terms of the summation are completely independent which is not strictly true.

Zachariasen[103] in his paper on sign relationships applied a similar expression,

$$s(h) = s\{\textstyle\sum_{h'} s(h') s(h + h')\} \tag{5.27}$$

to the solution of the structure of metaboric acid. The signs of forty structure factors were obtained in terms of symbols by means of inequality relationships. The application of the sign relationship then yielded a unique set of signs for the 168 largest unitary structure factors.

Woolfson[104] demonstrates the use of this method by deriving signs for an a-axis projection of dicyclopentadenyl di-iron tetra-carbonyl.

In order to make use of the relations in equations (5.26) and (5.27) it is necessary first to make use of inequalities to determine signs, and as the complexity of a structure increases, fewer unitary structure factors have sufficient magnitude for inequality relationships to hold true. Consequently the less reliable expression in equation (5.25) must be used, although the efficiency of this procedure can be increased by the introduction of statistical assumptions that some fraction of the sign relationships will be correct.

The weakness of direct methods is now apparent; firstly valid relationships between sign-symbols must be obtained, and secondly the preliminary allocation of signs, from which many others are derived, if wrong can lead to a massive snowballing of errors. Two approaches have been made to overcome these problems, one procedure yields a multi-solution answer of several sets of structure factor signs. The correct one is obtained by Fourier synthesis, i.e. the most chemically reasonable electron density map is chosen as correct. The other procedure yields one final set of structure factor signs which may be right or wrong.

Historically, the first attempt at direct methods was made by Banerjee* in 1933. Advances were made with the publication of certain inequality-relationships by Harker and Kasper[95] in 1948, and major contributions to the development of the methods were made independently by Sayre[99], Cochran[105], and Zachariasen[103] in 1952. Karle and Karle[106] extended earlier work of Hauptman and Karle[107] systematically with their symbolic addition procedures, and at the present time direct methods have reached a point where they are as effective as, and complementary to, heavy-atom methods. In the discussion below the symbolic addition procedures of Karle and Karle will be considered, together with examples of the application of computers to the use of direct methods in a single-solution and multi-solution case.

It must be borne in mind when direct methods are used that the Fourier maps that are calculated using E's as coefficients rather than F's, i.e. E-maps, are not as easy to interpret as the usual Fourier maps. The main reasons for this are that the E's represent perfectly sharpened atoms and also have been selected for their usefulness in the mathematical relationships that are used. As a result, spurious peaks may appear, some peaks which should be present may not appear, and no importance attaches to the relative heights of the peaks that do appear.

Symbolic Addition Procedures

Karle and Karle[106, 108] describe a symbolic addition procedure for phase determination of both centro- and non-centrosymmetric crystals which developed from the work described in *ACA Monograph* 3. When probability concepts are made part of the sign determining procedure, relatively few correct signs need to be chosen initially, and the Fourier maps (E-maps) calculated from the larger normalised structure factors readily give the atomic positions in the unit cell.

Proc. Roy. Soc. 141, 188 (1933).

The earlier procedure described by Hauptmann and Karle[107] began by selecting a basic set of signs from probability formulae which used measured X-ray intensities. Additional probability formulae were then used to obtain further phase information. The later procedure described by Karle and Karle revised the logic of the phase determining procedure while retaining the probability formulae and concepts. Originally the sign of E_h was given by the sign of $\Sigma_1 + \Sigma_2 + \Sigma_3 + \Sigma_4$. Woolfson[109] converts and simplifies the expressions so that the sign of U_h is obtained, i.e.

$$\Sigma_1 = \varepsilon_3/4\varepsilon^{5/2}(U_{\frac{1}{2}h}^2 - \varepsilon)$$

$$\Sigma_2 = \varepsilon_3/2\varepsilon^{5/2}\sum_{h'} U_{h'}U_{h+h'}$$

$$\Sigma_3 = \varepsilon_4/4\varepsilon^{7/2}\sum_{h'} U_{h'}(U_{\frac{1}{2}(h+h')}^2 - \varepsilon)$$

$$\Sigma_4 = \varepsilon_5/8\varepsilon^{9/2}\sum_{h'} (U_{\frac{1}{2}h+h'}^2 - \varepsilon)(U_{h'}^2 - \varepsilon)$$

In these expressions $\varepsilon_r = \sum_{j=1}^{N} n_j^r$, ε is written for ε_2, and the U's correspond to the structure factors for point atoms of weights n_m (the definition of n_j has been given earlier).

In their more recent procedure a small properly chosen set of normalised structure factors are allocated signs or symbols where the signs are unknown. In this way the Σ_2 relationship may be applied to the very largest structure factors at the beginning, without first having to determine their signs.

Karle and Karle point out that the first application of symbols to phase determination was described by Gillis[110] in 1948, and that their symbolic addition procedure when applied to centrosymmetric crystals has several similarities to that of Zachariasen[103], i.e. the use of symbols and the restriction of attention to structure factors of larger magnitude.

One consequence of using symbols is that a large number of sets of signs can be obtained which are consistent with the phase relation, and the question arises as to what restrictions should be applied to the number of symbols used. Karle and Karle overcome the problem by using probability relationships. Zachariasen in his metaboric acid determination used phase information derived from inequality relationships.

The Karle and Karle procedure is based on the recognition that with the proper use of probability information the number of ambiguous sets of signs can be kept to a small number. The name *symbolic addition procedure* is derived from the use of symbols for phases, and the use of addition procedures involving particular combinations of phases. This procedure uses in the centrosymmetric case the Σ_2 relationship of the earlier *ACA Monograph* 3, i.e.

$$sE_h \approx s\Sigma E_k E_{h-k}$$

together with the probability function described by Woolfson[111] and Woolfson and Cochran[101],

$$P_+(h) \approx \tfrac{1}{2} + \tfrac{1}{2}\tanh \sigma_3 \sigma_2^{-3/2} \left| E_h \right| \sum_k E_k E_{h-k}$$

where

$$\sigma_n = \sum_{j=1}^{N} Z_j^n$$

and Z_j = the atomic number of the jth atom.

Further expressions are given for the non-centrosymmetric case and a description[112] of the structure determination of the alkaloid reserpine by Karle and Karle makes use of these expressions.

Procedures used in Phase Determination

Karle and Karle[106]

(a) Values of the normalised structure factor magnitudes are produced from the measured intensity data by means of the following expression:

$$\left| E_h \right|^2 = \left| F_h \right|^2 / \varepsilon \sum_{j=1}^{N} f_j^2(h)$$

where F_h = the structure factor magnitude

$\quad\quad f_j$ = the atomic scattering factor for the jth atom

$\quad\quad \varepsilon$ = a number that corrects for space group extinction

(b) The $\left| E_h \right|$ values are divided into the eight subgroups defined by the parities of the h, k, l indices. Each subgroup is listed in order of decreasing magnitude. Alternatively the $\left| E_h \right|$ values can be listed so that all the combinations of k and $h-k$ are obtained for a given value of h, i.e. a Σ_2 listing. With each k and $h-k$ pair is listed the value of the function $\sigma_3 \sigma_2^{-3/2}$ $\left| E_h E_k E_{h-k} \right|$ for use with the probability formula.

(c) An origin is specified by choosing a correct set of $\left| E_h \right|$ and allotting phases to them. Additional phase symbols are next allotted to $\left| E_k \right|$ of large magnitude which appear in many combinations as required by the relationship Σ_2, i.e. $sE_h \approx s\Sigma_k E_k E_{h-k}$.

(d) The application of the Σ_2 relationship proceeds as follows: Five to ten of the largest $\left| E_h \right|$ for each parity group are listed and signs which specify the origin, plus one sign symbol, are assigned to the structure amplitudes. The Σ_2 relationship is then used to define as many signs of the largest $\left| E_h \right|$ as possible in terms of the specified ones and any other newly determined ones.

(e) The next largest five to ten $|E_h|$ for each subgroup are added, and the procedure is continued. Where necessary new symbols are introduced in order to determine the signs of the larger $|E_h|$ values in terms of specified signs and unknown symbols. As the determination is extended to $|E_h|$ of smaller magnitudes, consideration is given to the probability estimates that are made.

Karle and Karle have found that the maximum number of symbols they needed to use was six, and these reduced to four or less at the completion of the phase determination. If there have been P unknown symbols used, then it would be necessary to calculate 2^P Fourier maps to be sure that the correct one is included. Although the figure 2^P can be reduced, by redefining some symbols in terms of others, using other phase determining formulae, and rejecting sets of phases which would produce Fourier maps with obviously incorrect features, the use of high speed computers makes the calculation of 2^P Fourier maps quite feasible. In general, ten to fifteen of the largest independent $|E_h|$ values per atom of the asymmetric unit are used in the three-dimensional Fourier synthesis.

It is pointed out that if all the E-maps are calculated, then any homometric structures will be found. A procedure is also described that can be applied to the solution of non-centrosymmetric structures.

F. R. Ahmed and S. R. Hall[113]

A set of programs is described which can be applied to centrosymmetric crystals. The procedure is based on the Σ_2 relationship and the probability expression used by Karle and Karle. The programs are written in Fortran for the IBM 360 system, and broken up into five stages as described below:

Stage 1. The overall temperature factor and scale are calculated by a Wilson plot method.

Stage 2. Normalised structure factors and their statistics are calculated and sorted in order of magnitude.

Stage 3. Triplets of reflections are searched for those that are greater than a specified minimum magnitude and that satisfy the Σ_2 relationship.

Stage 4a. Origin defining reflections are chosen, symbols are assigned, and successive applications of the Σ_2 relationship are made to $|E_h|$ values greater than a specified optimum value.

Stage 4b. The signs of normalised structure factors having a magnitude which lies within certain limits less stringent than in stage 4a, are determined by direct application of the signs found in stage 4a to sets of equivalent pairs E_k, E_{h-k} related to each E_h. The structure factors are then sorted in preparation for a Fourier synthesis.

The procedure that is followed to minimise incorrect allocations of phases will be described in more detail, as it is interesting to compare it with that of Woolfson and Germain that will be described shortly.

The Σ_2 relationship can be written $s(E_1) \approx s(\Sigma E_2 . E_3)$, and stage 3 of the above programs contains three options in the search for reflection sets related by this expression:

1. A full search can be made of all E values for reflection sets E_1, E_2, E_3 with magnitudes greater than a specified minimum.
2. A partial search can be made for reflection sets having $E_1 \geqslant$ a minimum value, and E_2 and $E_3 \geqslant$ a specified intermediate value which is chosen to give a sufficient number of statistically preferable $E_2 E_3$ pairs related to each E_1.
3. A non-redundant partial search can be made which is similar to 2, but $E_1 > E_2 > E_3$. This ensures that each triplet appears only once in the listed output.

Usually a non-redundant partial search is sufficient for most structural problems, and less time consuming. Stages 4a and 4b of the programs then estimate the phases for structure factors which have magnitudes greater than the specified intermediate value and less than this value respectively. In this way successive applications of the Σ_2 relationship alone, which is a statistically uncertain procedure, are not employed; instead the signs of the largest structure factors are estimated using single E-triplets, and at a later stage these signs are applied to triplet sets associated with smaller E's to determine the signs of these lower magnitude structure factors.

The sign determining procedure of stage 4a is split into two parts, tentative allocation of signs and final allocation of signs. The first part allocates sixteen separate counters to each reflection for the accumulation of the sums $(\Sigma_k \sigma_3 \sigma_2^{-3/2} |E_h| E_k E_{h-k})$ for a maximum of 400 reflections. The acceptance of an estimated sign occurs when the corresponding sum is greater than a specified figure, which is usually set high initially and lowered as the sign allocation continues. The program proceeds by calculating the sign of a reflection from two other signed reflections in the triplet, i.e. the sign = product of the known signs. This information is added to the accumulating total for that particular reflection, and the sign is accepted when the total passes a specified test limit.

Reflections with acceptable tentative signs are then used to determine more signs where possible. When no more signs can be estimated the test limit is dropped to the next lowest value and the procedure is repeated. At certain points in the process of successively lowering the test limit, the allocation of sign symbols A, B, C or D is made when reflections still exist with undetermined signs. A symbol is allocated to the strongest unsigned reflection having the largest number of triplets. The sign allocation procedure is then repeated using the new test limit value. The second part of stage 4a is the determination of the signs of the sign symbols from a comparison of the accumulated sums for each reflection. The final sign for each reflection is determined by substituting for the symbols and adding the sums:

$$\left(\sum_k \sigma_3 \sigma_2^{-3/2} \, |\, E_h\, |\, E_k E_{h-k}\right)$$

that have been accumulated. The final sign is accepted when the total sum is above the minimum acceptable value set by the user.

Stage 4b extends the above procedure to lower magnitude structure factors and obtains a set of sorted reflections ready for input to an initial Fourier synthesis.

Woolfson and Germain[114]

A procedure is described for applying the symbolic addition method to centrosymmetric structures in such a way as to reduce the chances of allocating incorrect phases to structure factors both during the initial steps, when single sign relationships must be used, and also later in the acceptance of sign relationships that develop between sign symbols. In addition, there are descriptions of procedures for applying phase-determining techniques to non-centrosymmetric structures. In the present discussion only the centrosymmetric case will be considered, where a multi-solution procedure is described that is applicable to structures having up to 400 atoms per unit cell.

The authors consider first the initial stages of sign allocation which are based on single sign relationships where new signs are usually obtained in terms of symbols. At this point a typical magnitude of the product of three structure factors would be about 20, and the probability of the sign relationship between the structure factors being correct is given by

$$P_{max} = \tfrac{1}{2} + \tfrac{1}{2} \tanh \left(\frac{20}{N^{\frac{1}{2}}} \right)$$

where there are N equal atoms per unit cell.

If values of N are substituted in the above expression the probability decreases from a value of 0·999(7) when $N = 25$, to 0·881 when $N = 400$. It is seen, therefore, that as the structural complexity increases the probability of a single triple-product sign relationship (t.p.s.r.) being correct decreases.

Next is considered the probability of no error occurring when the first ten steps of the sign-allocation procedure are taken using single t.p.s.r.'s. For $N = 25$ the probability = 0·998, but when $N = 400$ the probability falls to 0·282. If, however, one allows for one error during the first ten steps then when $N = 25$ the probability of completing ten steps with one or less errors = 1·000, but when $N = 400$ the probability is 0·663. If this is extended to the probability of completing the first ten steps with two or less errors, then for $N = 25$ the probability is 1·000 and when $N = 400$ the probability = 0·894.

Therefore, if an allowance can be made for the possibility of making one or two incorrect sign allocations in the first ten steps, a method of progress will be obtained giving increased protection against failure. The process of allowing for one error in the first ten steps is as follows:

1. Assume the first step succeeds or fails.
2. If it fails assume the next nine steps succeed.
3. If it succeeds allow the second relationship to either fail or succeed, and so on.

This leads to eleven different routes, one corresponds to no failures and ten correspond to one failure.

In the cases where the number of steps taken using single t.p.s.r.'s is larger than ten, the number of routes which must be considered does not become prohibitively large. For twenty such steps when $N = 400$, making allowance for the probability of two or less failures, there will be 211 routes with a probability of 0·569 that one of the routes is correct. If $N = 400$ and twenty steps are taken, allowing for three failures there will be 1351 routes and a 0·723 chance of obtaining the correct one.

Once the initial stages of sign allocation have been passed and signs are allocated on the basis of two or more t.p.s.r.'s, then less chance of making incorrect sign allocations exists, e.g. the overall probability P_+ of a sign being correctly given by two weak sign relationships of probabilities P_1 and P_2 is given by:

$$P_+ = \frac{P_1 P_2}{1 - P_1 - P_2 + 2P_1 P_2}$$

if $P_1 = P_2 = 0.80$, then $P_+ = 0.941$, i.e. a good chance that the allocation is correct.

The authors then describe the following program which is written in Fortran IV for an IBM 360/40 computer with a 16K store. The five parts can be run together or separately.

Part I   Up to 250 $|U|$ or $|E|$ values are read in together with up to 1000 t.p.s.r.'s produced by another program from the 250 largest structure amplitudes. The program fixes the unit cell origin and allocates symbols to six other reflections that occur most frequently in the t.p.s.r.'s.

Part II   The relationship $s(\boldsymbol{h}) \approx s\{\Sigma_{\boldsymbol{h}'} s(\boldsymbol{h}')s(\boldsymbol{h}-\boldsymbol{h}')\}$ is used to build up signs and symbol allocations.

Part III   A search is made for relationships between symbols which are listed in order of probability decided by frequency of occurrence.

Part IV   The relationships between the symbols are solved using a method similar to that of Cochran and Douglas; e.g. in the simplest case, if a set of six strong non-linearly dependent relationships can be found and are assumed to be correct, then a unique set of signs can be allocated to the symbols. Allowance is also made for the possibility of one of the relationships failing, giving six more sets of signs for the symbols, i.e. seven in all. If all the symbols cannot be solved then the number of sets of signs for the symbols, allowing for one failure, is given by $2^{6-m}(m+1)$ where $m =$ the number of symbols for which a solution has been obtained.

Part V   The sign of each reflection is generated for each of the sets of signs for the symbols. If a reflection has more than one set of symbols indicating its sign and these are in opposition and the overall indication of sign is unsatisfactory, then the sign of this reflection is classed as indeterminate.

Three figures of merit are generated for each output set of signs:

(a) A count of the number of sign relationships which hold is given.

      (b) The sum of the probabilities of the sign relationships which hold is given.

      (c) The value $\Sigma s_1 s_2 s_3 \mid E_1 E_2 E_3 \mid$ is given.

The above program has been used successfully on a number of structures, e.g. Rabinowitz and Kraut[115] solved the structure of myo-inositol which had 96 atoms in the unit cell, space group $P2_1/c$. Of the 250 E values, the signs of 240 were correctly determined, eight were undetermined as they did not occur in t.p.s.r.'s, and two were determined incorrectly.

The Determination of Structure Factor Phases in the Non-centrosymmetric Case

J. Karle and I. L. Karle* in a paper on the symbolic addition procedure describe relationships that are valid in determining structure factor phases for both centrosymmetric and non-centrosymmetric crystals. In the latter case three relationships are presented after a detailed analysis involving both algebraic considerations and probability theory, together with many references to earlier papers.

In the non-centrosymmetric case the phase determining formulae are:

$$\phi_h \approx \langle \phi_k + \phi_{h-k} \rangle_{k_r} \tag{5.28}$$

$$\phi_h \approx \frac{\sum_{k_r} \mid E_k E_{h-k} \mid (\phi_k + \phi_{h-k})}{E_{k_r} \mid E_k E_{h-k} \mid} \tag{5.29}$$

and

$$\tan \frac{E_k \mid E_k E_{h-k} \mid \sin (\phi_k + \phi_{h-k})}{E_k \mid E_k E_{h-k} \mid \cos (\phi_k + \phi_{h-k})} \tag{5.30}$$

ϕ_h is the phase of the normalised structure factor E_h, and k_r designates the p terms in a Taylor expansion of the sine function; it implies that k ranges only over those vectors associated with large $\mid E \mid$ values.

Although a detailed derivation of these expressions cannot be given here it is instructive to consider the following much shortened description of how they were obtained.

Karle and Hauptman† have shown that the Harker-Kasper inequalities arise from the *positivity criterion*, i.e. the electron density distribution in a structure must be a non-negative function. As a result a complete set of inequalities could be derived which were

*Acta Cryst., **21,** 849 (1966).
†Acta Cryst., **3,** 181 (1950).

valid for all space groups. These may be written as a set of increasingly complex relationships:

$$F_{(000)} \geqslant 0 \tag{5.31}$$

$$|F_{(hkl)}| \leqslant F_{(000)} \tag{5.32}$$

$$|F_h - \delta| \leqslant r \tag{5.33}$$

where $\delta = \delta(h, k) = F_{h-k}F_k / F_{(000)}$
and

$$r = \left| \begin{matrix} F_{(000)} & F^*_{h-k} \\ F_{h-k} & F_{(000)} \end{matrix} \right|^{\frac{1}{2}} \left| \begin{matrix} F_{(000)} & F^*_k \\ F_k & F_{(000)} \end{matrix} \right|^{\frac{1}{2}} \Big/ F_{(000)}$$

also $h = h_1 + h_2, \quad k = h_2$

Equation (5.33) implies that the complex structure factor F_h is bounded by a circle in the complex plane whose centre is δ and radius r. It follows that F_h should be proportional to an average over the various $\delta(h, k)$ involving the larger $|F|$ as k is varied,

$$F_h \propto \langle F_k F_{h-k} \rangle_k$$

If only the larger structure factor magnitudes are considered and

$$F_h = |F_h| \exp(i\phi_h)$$

then approximately

$$\phi_h \approx \langle \phi_k + \phi_{h-k} \rangle_k$$

where ϕ_k and ϕ_{h-k} are associated with the complex number $\delta(h, k)$ through the relationship

$$\delta(h, k) = |\delta(h, k)| \exp[i(\phi_k + \phi_{h-k})]$$

In the original paper the derivation of equations (5.32) and (5.33) is carried out using quasi-normalised structure factors defined* as

$$\mathscr{E}_k = \sigma_2^{-\frac{1}{2}} \sum_{j-1}^{N} Z_j \exp(2\pi i\, k \cdot r_j)$$

where Z_j is the atomic number of the jth atom, having co-ordinates represented by the vector r_j in a unit cell containing N atoms

$$\sigma_n = \sum_{j=1}^{N} Z_j^n$$

In fact it is preferable to use normalised structure factors E_k.

Equations (5.32) and (5.33) come from a relation for $\mathscr{E}_h$ obtained from probability methods given by Karle and Hauptman† in an earlier paper:

$$\mathscr{E}_h \approx \sigma_2^{3/2} \sigma_3^{-1} \langle \mathscr{E}_k \mathscr{E}_{h-k} \rangle_k$$

*Karle and Hauptman, *Acta Cryst.*, **12**, 404 (1959).
†Karle and Hauptman, *Acta Cryst.*, **9**, 635 (1956).

This equation can be re-written,

$$1 \approx \sigma_2^{3/2}\sigma_3^{-1} \langle |\mathcal{E}_h^{-1}\mathcal{E}_k\mathcal{E}_{h-k}| \exp\left[i(-\phi_h+\phi_k+\phi_{h-k})\right]\rangle_k$$

If only those $|\mathcal{E}_k|$ and $|\mathcal{E}_{h-k}|$ having large values are used it can be shown that

$$\langle |\mathcal{E}_k\mathcal{E}_{h-k}| \sin(-\phi_h+\phi_k+\phi_{h-k})\rangle_{k_r} \approx 0$$

Expanding the sine function using Taylor's method then leads to an expression which may be rewritten and used as a phase determining formula, i.e. equation (5.29)

$$\phi_h \approx \frac{\sum_{k_r}|\mathcal{E}_k\mathcal{E}_{h-k}|(\phi_k+\phi_{h-k})}{\sum_{k_r}|\mathcal{E}_k\mathcal{E}_{h-k}|}$$

in which the ϕ_h are defined as linear functions of the phases. If all the $|\mathcal{E}|$ are roughly the same order of magnitude then

$$\phi_h \approx \langle \phi_k+\phi_{h-k}\rangle_{k_r}$$

in agreement with equation (5.28).

Equation 5.30 as already mentioned is also derived from the relation,

$$\mathcal{E}_h \approx \sigma_2^{3/2}\sigma_3^{-1}\langle \mathcal{E}_k\mathcal{E}_{h-k}\rangle_k$$

If $\mathcal{E}_h = |\mathcal{E}_h|\cos\phi_h + i|\mathcal{E}_h|\sin\phi_h$ then

$$|\mathcal{E}_h|\cos\phi_h \approx \sigma_2^{3/2}\sigma_3^{-1}\langle |\mathcal{E}_k\mathcal{E}_{h-k}|\cos(\phi_k+\phi_{h-k})\rangle_k \quad (5.34)$$

and

$$|\mathcal{E}_h|\sin\phi_h \approx \sigma_2^{3/2}\sigma_3^{-1}\langle |\mathcal{E}_k\mathcal{E}_{h-k}|\sin(\phi_k+\phi_{h-k})\rangle_k \quad (5.35)$$

Dividing equation (5.34) by equation (5.35):

$$\tan\phi_h \frac{\langle |\mathcal{E}_k\mathcal{E}_{h-k}|\sin(\phi_k+\phi_{h-k})\rangle_k}{\langle |\mathcal{E}_k\mathcal{E}_{h-k}|\cos(\phi_k+\phi_{h-k})\rangle_k}$$

If E_k replace $\mathcal{E}_k$ the quasi-normalised structure factors then an expression is obtained which is analogous to equation (5.30), the 'tangent formula'.

A procedure that may be followed in applying the three phase determining formulae is as follows:

1. A list of corrected normalised structure factors is prepared as for a centrosymmetric case.
2. Listings are prepared for use with, in the first instance, equation

(5.28). In addition values of κ are listed with each pair k and $h-k$ for use with the probability formula*

$$P_k(\phi_h) \approx \left[2\pi I_0(\kappa)\right]^{-1} \exp\left[\kappa \cos\left(\phi_h - \phi_k - \phi_{h-k}\right)\right]$$

where I_0 is a Bessel function† and

$$\kappa = \kappa(h, k) = 2\sigma_3 \sigma_2^{-3/2} \left| \mathscr{E}_h \mathscr{E}_k \mathscr{E}_{h-k} \right|.$$

3. As well as specifying the origin (as for the centrosymmetric case) the enantiomorph is determined by the assignment of a sign to a particular linear combination of phases which satisfies the definition of an invariant‡. Symbols may also be assigned.

4. Equation (5.28) is now used to determine the phases of the remaining $|E_h|$ of large value in terms of phase specifications and unknown symbols.

5. An initial list of one to two hundred phases determined in terms of specified phases and unknown symbols may be listed and eventually added to by using the other two formulae, equations (5.29) and (5.30) after specific values are introduced for the unknown symbols.

6. Finally, E maps may be calculated with the largest $|E_h|$ values (about 30% of the data).

Karle and Karle point out that in addition to their usefulness as sign determining formulae, equations (5.29) and (5.30) by re-iteration can effect convergence of the phase values to a more accurate set of values. They also may be used to apply criteria for the rejection of poorly determined phases.

Examples of the use of direct methods in solving non-centro-symmetric structures are now fairly common in the literature. For example, the structure of 7-hydroxy-1,9-10-trimethoxy-4-azabicyclo 5,2,2 undeca-8,10-dien-3-one, which is a photolysis product of N-chloroacetylmexaline, was solved by Isabella L. and J. Karle§ using normalised structure factors and the 'sum of the angles' formula $\phi_h \approx \langle \phi_k + \phi_{h-k} \rangle_{k_r}$ together with the tangent formula

$$\tan \phi \approx \frac{\sum_k |E_k E_{h-k}| \sin(\phi_k + \phi_{h-k})}{\sum_k |E_k E_{h-k}| \cos(\phi_k + \phi_{h-k})}$$

---

*Cochran, W., *Acta Cryst.*, **8**, 473 (1955).

†Watson, G. N., *Theory of Bessel Functions*, Cambridge University Press (1945).

‡See for example Karle and Hauptman, *Acta Cryst.*, **9**, 45 (1956), Karle and Hauptman, *Acta Cryst.*, **14**, 217 (1961)

§*Acta Cryst.*, B26, **9**, 1276 (1970).

See also the crystal structure of solaphyllidine by Isabella L. Karle, *Acta Cryst.*, B26, 1639 (1970).

SOME OTHER APPROACHES TO THE PHASE PROBLEM

Isomorphous Replacement

The term *isomorphous* means different things to different people. Mitscherlich (1819) considered two substances to be isomorphous if the respective interfacial angles of the different crystals were identical or nearly so, e.g. KH_2PO_4 and KH_2AsO_4. If two different crystals displayed identical interfacial angles it was considered likely that the substances differed only in the substitution of one atom for another of similar chemical nature in the formula.

Isomorphous substances are nowadays considered to be those in which geometrically similar structural units are arranged in the same way in the crystal. In this way substances such as NaCl, CaF_2 and diamond, which can each display the same external form but which have nothing in common chemically, are not now considered to be isomorphous. Mathematical crystallographers treat isomorphism as a relationship between the space group of a crystal and the crystal class or point group.

In protein crystallography the term isomorphism is usually applied to heavy atom derivatives of the protein, i.e. isomorphous derivatives, and the technique of using these derivatives is an isomorphous replacement technique. The different heavy atoms that are introduced into the protein crystals do not necessarily occupy the same sites in each derivative.

The first structure to be solved using an isomorphous replacement technique was phthalocyanine* which is sufficiently isomorphous with its nickel derivative to allow the allocation of phases to be made. The unit cell dimensions and space group, $P2_1/a$, of each compound are identical but the metal derivative has a nickel atom at the centres of symmetry at 0,0,0, and $\frac{1}{2},\frac{1}{2},0$; whereas the organic molecule has two imide hydrogen atoms one each side of the centre of symmetry. A comparison of the absolute intensities of both structures showed that the intensities of some of the reflections from the nickel complex were greater than the corresponding reflections of the organic crystals and that some were less. In the former case the structure amplitude of the organic molecule will be positive and in

*Robertson, J. M., *J. Chem. Soc.*, 1195 (1936).

the latter case negative. In this way the signs of 300 of the structure amplitudes for the phthalocyanine structure could be allocated.

More commonly isomorphous replacement techniques may be applied to metal complexes where suitable pairs of metals can be used. Usually two isomorphous complexes are made each containing a different metal ion of a similar size, for example potassium and rubidium or magnesium and zinc can sometimes be used to obtain isomorphous derivatives.

Isomorphous replacement methods have been particularly useful in the elucidation of protein structures and the first example in this field was the structure determination of haemoglobin by Perutz *et al.** The isomorphous mercury derivative of haemoglobin was prepared by the reaction of the sulph-hydryl groups of the protein with *p*-chloromercuribenzoate. The product was a haemoglobin molecule containing two mercury atoms bound at specific sites, having identical cell dimensions with the original protein. A comparison of the diffraction patterns of both compounds showed clear differences in certain reflections. A Patterson projection down the *b*-axis using $|\Delta F|^2$ as coefficients led to the determination of the co-ordinates of the mercury atoms. The signs of the structure amplitudes of the protein were found as for phthalocyanine by comparing the magnitudes of the structure amplitudes in the protein with similar ones in the metal-protein derivative. A projection of the electron density down the *b*-axis could then be calculated, and this led ultimately to a view of a single layer of haemoglobin molecules.

Similar studies have been applied to horse haemoglobin where both the silver and the mercury derivatives of the protein have been made. Mercuri-iodide and auri-chloride have also been used to produce isomorphous derivatives of ox-haemoglobin.

In the case of myoglobin alternative procedures had to be found to prepare isomorphous derivatives†. Myoglobin was crystallised in the presence of various inorganic ions containing heavy elements, e.g. HgI_4^{2-}. X-ray photographs were then taken to show that intensity differences occurred between the diffraction pattern of the protein and that of the hoped for metal-protein derivative. Combination of the metal ion was taken to be at a specific site in the protein when a difference Patterson map showed only one peak per asymmetric unit.

Other heavy atom derivatives of proteins have been prepared using platinum, uranium, and cadmium as marker atoms.

---

*Green, D. W., Ingram, V. M., Perutz, M. F., *Proc. Roy. Soc.*, A225, 287 (1954).

†Crick, F. H. C. and Kendrew, J. C., *Advances in Protein Chemistry*, Vol. XII, p. 189, Academic Press (1957).

A measure of the suitability of an isomorphous derivative is given by the expression*

$$M_h \sqrt{\left(\frac{n}{M_p}\right)}$$

where n = number of heavy atoms per protein molecule
M_h = atomic weight of the heavy atom
M_p = molecular weight of the protein.

Sometimes the heavy atom may only have a site occupancy of perhaps 80% of the possible sites and allowance may be made for this as follows:

$$k(M_h - 25)\sqrt{\left(\frac{n}{M_p}\right)}$$

where k is the occupancy of the heavy atom ($= 1$ for full occupation). Using this expression for real structure factors, a value of $1 \cdot 0$ is satisfactory and $0 \cdot 5$ is marginal; complex structure factors need higher values, $1 \cdot 5$–$2 \cdot 0$.

An alternative approach to isomorphous replacement methods is to use a substance containing a suitable marker atom, and take X-ray photographs using two wavelengths, one near the absorption edge of the marker atom. The scattering power of this atom will usually be reduced as a result of anomalous scattering. The same information is thus obtained as in the isomorphous replacement method, with the added advantage that it is perfect isomorphism as only the one substance is being used. If the wavelength used is shorter than that of the absorption edge the phase shift of the wave from the anomalously scattering atom must be allowed for. If the wavelength is longer, then there is no phase shift only a reduction in scattering power.

Anomalous Scattering

In the case where the incident X-ray beam wavelength lies close to the absorption edge of an atom, the atomic scattering factor, f, was described earlier as

$$f = f_0 + f' + \mathrm{i}f''$$

where f' and f'' are correction factors for the real and imaginary parts of the anomalous scattering. Use may be made of the pheno-

---

*Crick, F. H. C. and Kendrew, J. C., *Advances in Protein Chemistry*, Vol. XII, p. 193, Academic Press (1957).

nena of anomalous scattering (or dispersion) in structure analysis
nd the following table gives a classification of the effects:

Phase determination by anomalous scattering (A.S.)

(a) A.S. without phase change $f'' = 0$, $f' \neq 0$. f' is almost independent of scattering direction (perfect isomorphism)

(b) A.S. with phase change $f'' \neq 0$*†

(c) Centrosymmetric structures‡§

(d) Non-centrosymmetric structures‖¶

(a) A.S. without phase change may be used to differentiate
between like atoms in a structure in those cases where such a
differentiation is not possible by prior chemical or geometrical
knowledge. A suitable radiation is used which will produce anoma-
ous scattering in one of the types of atoms but not in the other**.

(b) The determination of the absolute configuration of a compound
capable of existing in two enantiomorphous forms is biologically
important. Anomalous scattering with phase change was used to
determine the absolute configuration of sodium rubidium tar-
rate††‡‡. (It can also be used to determine the phases of diffracted
rays from non-centrosymmetric crystals§§‖‖.) The structure of the
compound was known prior to the determination of the absolute
configuration. Significant differences were observed in the intensities
of some hkl and $\overline{hkl}$ reflections using ZrK_α radiation. As the atomic
co-ordinates of the structure were known, the structure amplitudes
could be calculated for the two enantiomorphous configurations;

*Peerdeman, A. F., Bommel, A. J. V. and Bijfoet, J. M., *Proc. Acad. Sci. Amst.*, **54**, 16 (1951); *Nature*, **165**, 271 (1951).
†Ramachandran, G. N. and Raman, S., *Current Sci. (India)*, **25**, 348 (1956).
‡Lipson, H. and Cochran, W., *Crystalline State*, Vol. III p. 227, Bell (1953).
§Ramaseshan, S., Venkatesan. K., and Mani, N. V. *Proc. Indian Acad. Sci.*, A46, 95 (1957).
‖Ramaseshan, S., and Venkatesan, K., *Current Sci. (India)*, **26**, 352 (1957).
¶Pepinsky, R. and Okaya, Y., *Proc. Nat. Acad. Sci., Wash.*, **42**, 286 (1956); *Phys. Rev.*, **103**, 1645 (1956).
Mark, H. and Szillard, L., *Z. Phys.*, **33, 688 (1925).
††Bijfoet, J. M., *Proc. Acad., Sci., Amst.*, **52**, 313 (1949).
‡‡Peerdeman, A. F., Bommel, A. J. V., Bijfoet, J. M., *Proc. Acad. Sci., Amst.*, **54**, 16 (1951); *Nature*, **165**, 271 (1951).
§§Bokhoven, C., Schoone, J. C., Bijfoet, J. M., *Acta. Cryst.*, **4**, 275 (1951).
‖‖ Ramachandran, G. N., and Raman, S., *Current Sci (India)*, **25**, 348 (1956).

a comparison could then be made with the calculated and observed intensities to decide whether I_{hkl} was greater or less than $I_{\bar{h}\bar{k}\bar{l}}$ for a particular configuration, and the absolute configuration could then be determined. Let us consider a reciprocal lattice point P_{hkl} cutting the sphere of reflection (Figure 5.16). This implies that the set of planes hkl are in a position to reflect the primary X-ray beam AB is a plane of the set hkl. It can be seen that the reflection hkl is the same as the reflection $\bar{h}\bar{k}\bar{l}$, differing only in being reflected from the opposite side of the planes hkl. The intensities of these reflections are such that $I_{hkl} = I_{\bar{h}\bar{k}\bar{l}}$ according to Friedel's law.

The breakdown of Friedel's law occurs where anomalous scattering with phase change occurs, i.e. $f'' \neq 0$. When $f' \neq 0$ and $f'' = 0$

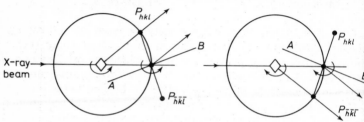

Figure 5.16. Reciprocal lattice point cutting the sphere of reflection

i.e. anomalous scattering without phase change, there is no breakdown in Friedel's law.

Under normal scattering conditions a pair of reflections hkl and $\bar{h}\bar{k}\bar{l}$ would have equal intensity assuming that the necessary corrections for absorption, etc. have been applied. When anomalous scattering is present $I_{hkl} \neq I_{\bar{h}\bar{k}\bar{l}}$, and such a pair of reflections are known as a Bijfoet pair. It is the comparison of such Bijfoet pairs that leads to a knowledge of absolute configurations. Under normal scattering conditions, the equivalence of reflections may be inferred from the Laue group symmetry elements. It is the point group symmetry which governs the equivalence of reflections when anomalous scattering occurs. The difference between I_{hkl} and $I_{\bar{h}\bar{k}\bar{l}}$ may also be used to determine the phases of diffracted rays in noncentrosymmetric structures*.

(c) Anomalous scattering without phase change corresponds to perfect isomorphism, since the use of a particular X-ray wavelength to cause anomalous scattering by a set of atoms in a structure corresponds to replacing that set of atoms by another set of different scattering power.

*Ramaseshan, S., Venkatesan, K., Mani, N. V., *Proc. Indian Acad. Sci.*, A46, 95 (1957).

The procedure that is followed is to use the small differences observed in the diffracted rays intensities in the case where normal scattering occurs, and in the case where anomalous scattering occurs. An example of the use of this technique in the centrosymmetric case is the solution of the structure of $KMnO_4$* by Ramaseshan and co-workers. In a description of the procedure that was followed† Ramaseshan points out the importance of applying the proper corrections for geometrical and physical factors. For example, if the crystal that is used is of an irregular shape, then absorption corrections must be applied to allow for the attenuation of incident and reflected beams within the crystal to a different extent dependent upon the path traversed.

(d) The procedures that are followed in (c) above and in (d) are all those used in the application of isomorphous replacement techniques. However, certain problems exist when the non-centrosymmetric case is considered, which may be overcome by using two incident wavelengths separately, one of which causes some atoms in the structure to scatter anomalously, the other of which produces normal scattering by all the atoms. Two types in this class may be differentiated (i) the anomalous scatterer lies on a centre of symmetry i.e. its phase $= 0$, and (ii) the anomalous scatterer does not lie on a centre of symmetry i.e. its phase $\neq 0$.

There is a difficulty in assigning the correct phase angle to the reflections which may be overcome by using both possible phases in the calculation of a Fourier synthesis. In the case of (i) this leads to a map containing a spurious centre of symmetry and in the case of (ii) the incorrect components are absorbed into the general background‡.

In all the applications of anomalous scattering to structural problems the procedure to be followed is in some measure dictated by the particular problem, and modifications are often necessary. It may be advantageous, for example, to use more than one set of atoms in a structure as anomalous scatterers (if this is possible). This of course would lead to an increase in the number of kinds of radiation used for the collection of intensity data.

Structure Factors

Anomalous scattering

If we consider two enantiomorphous compounds A and B in space

*Ramaseshan, S., Venkatesan, K., Mani, N. V., *Proc. Indian Acad. Sci.*, A46, 95 (1957).
†*Advanced Methods of Crystallography*, Academic Press, p. 84(1964).
‡Ramachandran, G. N., Raman, S., *Acta Cryst.*, **12**, 957 (1959) (See also Ramachandran, G. N., *Advanced Methods of Crystallography*, p. 46 Academic Press (1964)).

group $P1$, their structure factors are given by

$$F_A(hkl) = \mid F(hkl) \mid \exp(i\alpha)$$

$$F_B(hkl) = \mid F(hkl) \mid \exp(-i\alpha)$$

The structure factors of the $\overline{hkl}$ reflections are given by

$$F_A(\overline{hkl}) = \mid F(hkl) \mid \exp(-i\alpha)$$

$$F_B(\overline{hkl}) = \mid F(hkl) \mid \exp(i\alpha)$$

It follows that with normal scattering

$$\mid F_A(hkl) \mid = \mid F_B(hkl) \mid = \mid F_A(\overline{hkl}) \mid = \mid F_B(\overline{hkl}) \mid$$

and

$$I_A(hkl) = I_A(\overline{hkl}) = I_B(hkl) = I_B(\overline{hkl})$$

i.e. it is impossible to differentiate between two enantiomorphous structures using normal scattering methods.

Anomalous scattering with phase change usually occurs so that f' is negative, and f'' is ahead of the real part $f_0 + f'$ by $\pi/2$ (see earlier the expression $f = f_0 + f' + f''$). This results in the breakdown of Friedel's law and

$$\mid F(hkl) \mid \neq F(\overline{hkl})$$

and

$$I(hkl) \neq I(\overline{hkl})$$

This change can be seen in Figure 5.17. The amplitude of the hkl reflections in (b) is the same as that of the $\overline{hkl}$ in (d) if no phase change occurs. However when phase change does occur in the diffracted ray from one of the sets of atoms, the phase of the wave from those atoms is advanced and in one case weakened and in the other case strengthened. The two reflected waves (shown dotted) no longer have the same amplitude.

The effect is one of shifting the two waves in (a) and (c) so that they lie in one case more in phase giving an increase in amplitude, and in the other case more out of phase giving a corresponding reduction in amplitude.

Isomorphous Replacement – The Centrosymmetric and Non-Centrosymmetric Cases

Two isomorphous structures will have structure factors identical in phase but differing in amplitude since the contribution of the heavy atom in each structure will differ. The contribution from the rest of the structure will be the same.

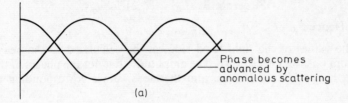

Phase becomes
advanced by
anomalous scattering

(a)

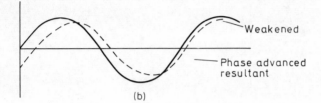

Weakened

Phase advanced
resultant

(b)

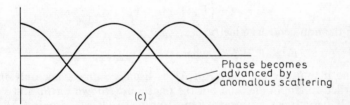

Phase becomes
advanced by
anomalous scattering

(c)

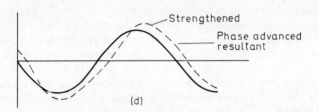

Strengthened

Phase advanced
resultant

(d)

Figure 5.17. Anomalous scattering with phase change. (a) Reflection hkl from two different sets of atoms in a structure. (b) Resultant of (a) and phase advanced resultant. (c) Reflection h̄k̄l̄. (d) Resultant of (c) and phase advanced resultant

Thus we have

$$\text{Structure A} \quad F_A = F_{rest} + F_{heavy\,1}$$

$$\text{Structure B} \quad F_B = F_{rest} + F_{heavy\,2}$$

Hence $F_A - F_B = F_{heavy\,1} - F_{heavy\,2}$

The values of the right-hand side can be calculated as the heavy atom positions are known. The amplitudes but not the phases of the left-hand side are known, and therefore values corresponding to

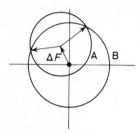

Figure 5.18. Argand diagram

each of the following four combinations are used in a trial and error procedure to find the ones that give the best agreement:

$$(\pm\,|\,F_A\,|) - (\pm\,|\,F_B\,|)$$

In the non-centrosymmetric case

$$F_B = F_A + \Delta F \quad \text{where } \Delta F = F_{heavy\,2} - F_{heavy\,1}$$

The magnitude and phase of ΔF is usually known, but only the structure amplitudes and not the phases are known for F_B and F_A.

If two circles are drawn representing the locus of points $|\,F_B\,|$ from the origin, and the locus $|\,F_A\,|$ from the vector ΔF, then the intersection of the two circles gives the points for which the above expression is valid and determines the phases of F_A and F_B. This Argand diagram (see Figure 5.18 and text below) yields two solutions as described earlier and various procedures may be applied to resolve which of the two is correct, e.g. use two sets of isomorphous replacements, use a combination of isomorphous and heavy atom methods, or use a Fourier synthesis with modified coefficients including both possible phases for each reflection.

Argand Diagrams

Argand diagrams may be constructed to derive the phase of the structure factor of a set of light atoms, when the positions and scattering powers of the heavy atoms of an isomorphous derivative are known. The structure factor for the heavy atoms is calculated and the moduli $|F_{\text{heavy}}|$ and $|F_{\text{heavy}+\text{light}}|$ are the experimentally measured values.

Two different orientations of the phase diagram are obtained which in general lead to two values of ϕ the phase of the light atom structure factor. The use of a second heavy atom derivative will then theoretically lead to an unambiguous allocation of phase [i.e. point A in Figure 5.19(c) where $|F_{LH_1}|$ and $|F_{LH_2}|$ coincide]. 0_L, 0_{LH_1} and 0_{LH_2} are the centres of the circles of respective radii $|F_L|$, $|F_{LH_1}|$ and $|F_{LH_2}|$. The circles 0_L and 0_{LH_1} intercept at A and B giving ambiguity in the value of ϕ, the phase angle of the light-atom structure factor. The circle 0_{LH_2} also intercepts the circles 0_L, 0_{LH_1} at A, so fixing the phase of F_L.

In practice more than two derivatives are usually needed and phases fixed in this way seldom produce circles meeting at a point. Errors arising from incorrect heavy atom positions, poor structure amplitude data, or poor isomorphism can all be treated as errors

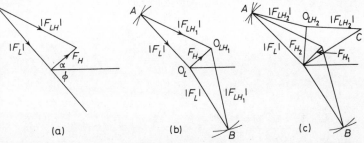

Figure 5.19. Argand diagram to derive the phase of the structure factor of a set of light atoms. (a) Construction of a diagram where F_H (i.e. phase α and magnitude $|F_H|$ is known. $|F_L|$ and $|F_{LH}|$ are obtained from the measured intensities. (b) The ambiguity in finding ϕ when only one heavy atom derivative (structure-amplitude $|F_{LH_1}|$) is used. (c) Resolution of the ambiguity by using a second heavy-atom derivative (structure amplitude $|F_{LH_2}|$)

in $|F_H|$ or $|F_{LH}|$. A function $P(\phi)$ can be used to describe the probability distribution of ϕ the phase angle, and the Fourier synthesis yielding the lowest mean-square error in electron density is that using phases chosen from the centre of gravity of the $P(\phi)$ distribution*. (See for example Cullis, A. F., Muirhead, H., Perutz, M. F., Rossmann, M. G., and North, A. C. T., *Proc. Roy. Soc.*, A265, 15 (1961); and Holmes, K. C., *Computing Methods in Crystallography* p. 183, Ed. J. S. Rollett, Pergamon Press).

*Blow, D. M., Crick, F. H. C., *Acta. Cryst.*, **12**, 794 (1959).

An alternative method to using three heavy-atom derivatives is to use two and treat the data by means of a Fourier synthesis in which both possible phases are taken into account for each reflection*. The correct phases should then reinforce one another leading to peaks of electron density at atomic sites. The incorrect phases are lost in the background.

---

*Kartha, G., *Acta Cryst.*, **14**, 680 (1961); Blow, D. M., Rossmann, M. G., *Acta Cryst.*, **14**, 1195 (1961).

Part 4
Crystallographic Computer Programs

6 Descriptions of Some Programs used in Crystallography

INTRODUCTION

The list of programs which are described is not intended to be comprehensive. It includes the more common programs that would be used in solving crystal structures, and this chapter is intended to give the reader some idea of the programs that are available and of the data that are needed to run them. When a library of crystallographic programs is available to the user he will obviously have access to more specialised information about the requirements of particular programs designed for specific computers. If the user intends writing his own programs then this list, together with the information in Part II, will provide a useful framework within which to work.

A set of data which can be used for testing standard crystallographic programs is to be made available by the International Union of Crystallography Commission on Crystallographic Computing, and should be included in Volumes II and III of *International Tables for X-ray Crystallography* as a supplement.

The approach to crystallographic computing is largely decided by the facilities available. Ideally a large computer with a store in excess of 16K is needed, so that a crystallographic library can be stored on disk or tape and the programs run as required. Once the intensity data have been collected, they too can be stored and from this point onwards everything can be controlled by the computer. This ideal situation does not always apply, and often the time available on a large computer is limited, although a smaller computer (8K) may be readily available.

There are two possible approaches in this case. For example, with an IBM 1130 computer, having 16K store and a multi-disk drive unit, programs designed for larger machines can be broken down and run piecemeal, the governing factor probably being the time needed. Alternatively, it is possible to run many of the preliminary programs, such as intensity corrections, Patterson projections, Harker sections and Fourier sections, on a machine with an 8K store and then transfer to a larger computer, e.g. the one at the S.R.C. Atlas Computing Laboratory at Didcot, for the later stages of the structure solution such as least-squares refinement and three-dimensional difference-Fourier syntheses.

The *World List of Crystallographic Computer Programs* (2nd edition), published by the International Union of Crystallography, contains listings of 697 programs written for a variety of computers including the English Electric KDF 9, the Ferranti Atlas, and the IBM 360. The languages used are in the main FORTRAN or ALGOL.

Two crystallographic program systems are described in the list; ATSYS used at Imperial College, London, and X-ray 63 which, recently up-dated, is available at the S.R.C. Atlas Computing Laboratory at Chilton, Didcot, Berks.

A set of crystallographic programs written in Fortran IV for use on the IBM 360 system has been produced by Dr. F. R. Ahmed at the Division of Biochemistry and Molecular Biology of the National Research Council, Ottawa 7, Canada.

Dr. M. Laing at the University of Natal has prepared a set of programs that may be run on an IBM 1130 computer[116].

INTENSITY CORRECTION PROGRAM

Purpose. This program is used to apply Lorentz-polarisation correction factors to the raw data obtained from films. A Philips spot-shape correction can also be applied, and the facility to correct for attenuation of diffracted rays by absorption in the crystal is usually included for the cases where the crystal can be approximated to a sphere or a cylinder.

Data needed

(a) A set of raw intensity data in a suitable format, i.e. three indices describing each plane together with the value of the measured intensity associated with it.

(b) The cell parameters a, b, c and α, β, γ, or the reciprocal values.

(c) If the reciprocal cell values are given in dimensionless units, then λ, the wavelength of the radiation, is needed.

(d) The angle μ for each layer of Weissenberg data.

(e) The linear absorption coefficient, together with the dimensions of the sphere or cylinder.

Output. It is usual to have both a printed output and an output on cards or tape which is in a suitable format to be used as input for later programs. Alternatively this output can be stored on magnetic tape or disk for later use. Values of both $|F_0|^2$ and $|F_0|$ are needed as output, the former being used in the calculation of Patterson syntheses and latter in the calculation of Fourier syntheses. A useful addition to this information is the value of $\sin \theta$ for each reflection. This then gives a built-in check that no reflection has a value of $\sin \theta > 1$, which may arise if the cell parameters have been wrongly calculated.

If the output is stored straight on to magnetic tape or disk, then there must also be a means of editing this data available so that duplicated reflections can be removed after averaging, and any other manipulation of the data that is necessary can be carried out.

DATA REDUCTION PROGRAM

Purpose. This program is used in the initial treatment of diffracto-meter (or film) data and can be divided into two parts:

Part 1 Diffractometer data usually consists of a set of indices and six other associated measurements. These are two background intensities, the spot intensities and the times taken for each of these measurements. Alternatively, the time of scan in seconds per degree and the angle of scan are known. Standard reflections are also included in the data at regular intervals as a check upon any deterioration in the crystal or instrument. Some reflections are so intense that attenuators are used in front of the counter, and their use must also be allowed for.

The first part of the data reduction program must therefore be able to:
  (a) Scale the data with reference to any attenuators that have been used.
  (b) Scale the data with reference to the standard reflections.
  (c) Average equivalent reflections and discard poor or unnecessary values.
  (d) Calculate the total background count and make allowance for this in the measured value of a re-flection.
  (e) Identify unobserved reflections in terms of a limit set by the operator, which is usually a function of the difference between measured intensity and back-ground intensity.

Part 2 The information obtained from Part 1 may now be cor-rected for Lorentz-polarisation factors, if these were not allowed for during the data collection. Absorption corrections may also be applied for the spherical or cylindrical crystal. The program then stores the data on magnetic tape as lists of standard format suitable for input to later programs. In addition, the following information may also be stored for later use:

(a) The unit cell parameters
(b) The space-group symmetry elements
(c) The form-factors
(d) Absorption coefficients and crystal dimensions.

The program should also calculate $\sin^2 \theta$ for each reflection which is used in the absorption correction, interpolate on the scattering factor curves, and derive $|F_0|$ from the measured intensities. Weights are allocated to each reflection based on counting statistics.

Data needed. Each reflection will have associated with it:

(a) Two background measurements.
(b) The time taken for these (or the rate of scan over a known angle).
(c) The reflection intensity measurement.
(d) The time taken (or the rate of scan over a known angle).
(e) The indices of the plane.

In addition, the following data will also be needed:

(a) The cell parameters and radiation wavelengths.
(b) Some indication of the equivalent reflections.
(c) The type of data collection used, i.e. film or counter, and if counter the mode of scanning (i.e. ω, 2θ or Weissenberg geometry). If Weissenberg geometry film data is used, then the layer angle μ will be needed.
(d) Whether or not attenuators have been used, and their values.
(e) The fraction of the background count below which the reflection is to be considered unobserved.
(f) A means of recognising background imbalance, i.e. the two background measurements should not differ too widely.
(g) Allocation of standard reflections.

Output. The printed output will contain:

(a) Indices of each plane.
(b) An observed or unobserved marker.
(c) The total count for each reflection.
(d) The background count for each reflection and whether the two background counts showed imbalance.
(e) The net count for each reflection.
(f) An allocated weight for each reflection.

If the output is not stored by the computer on magnetic tape or disk, then a similar output to the above is produced on cards or paper tape. Once this program has been run a permanent data record has been obtained which all the later programs will use.

PATTERSON SYNTHESIS PROGRAM

Usually one program is used for both Patterson and Fourier calculations. For clarity, each is described separately here.

Purpose. The function $P(UVW) = 1/V_c \sum_h \sum_k \sum_l |F_{(hkl)}|^2 \cos 2\pi (hU + kV + lW)$ where h, k, l are summed from $-\infty$ to $+\infty$ can be used to obtain the vector relationships between the scattering material in the unit cell. A Patterson synthesis program produces a three-dimensional contour map of the vector relationships between the atoms in a unit cell from the experimental values that have been obtained for $|F_{(hkl)}|^2$. The user specifies the intervals along each axis for which a value of $P(UVW)$ must be calculated, and the output is in the form of sections of vector space taken up a specified axis. The program may have a peak-scan procedure incorporated into it so that the co-ordinates of any peaks above a specified value will be printed on the output. Alternatively once the output has been obtained, it is necessary to draw in the contours on each section at any arbitrary chosen value, and examine each section visually for the co-ordinates of peaks. If the crystal system does not have orthogonal axes then it is an advantage to be able to obtain the output sections printed with respect to axes set at the correct oblique angle, so that contours can be drawn on the output without first having to transfer the output manually to an oblique set of axes.

Data needed.

(a) A set of $|F_{(hkl)}|^2$ values for each plane which have been sorted on each of the *hkl* indices, so that, for example, *h* may be the fastest moving index, *l* the slowest, and *k* the intermediate index.

(b) The intervals along each axis for which a calculation is done must be specified, e.g. 1/30 or 1/60 of the cell edge may be used.

(c) The axis up which the sections are to be taken must be specified.

(d) Usually a restriction is placed on the amount of output required so that only a unique volume of the vector space is obtained, i.e. duplication as a result of symmetry relationships is avoided.

(e) The angle between the axes on the output sections is specified for oblique crystal systems.

(f) Usually the expression used for calculating a Patterson synthesis is modified so that the trigonometric terms are products and not sums. (This is explained more fully in the case of the Fourier synthesis.) See p. 130 for Patterson formulae for each crystal system.

As an example, consider space group No. 2, $P\bar{1}$. If a projection is being calculated for example down the a-axis and $0kl$ and $0k\bar{l}$ data are being used, then:

$$P(YZ) = \sum\sum \{\,|\,F_{(0kl)}\,|^2\,(\cos 2\pi kV + \cos 2\pi lW) +$$

$$|\,F_{(0k\bar{l})}\,|^2\,(-\cos 2\pi lW + \cos 2\pi kV)\}$$

In terms of trigonometric products this becomes:

$$P(YZ) = \sum\sum \{(\,|\,F_{(0kl)}\,|^2 + |\,F_{(0k\bar{l})}\,|^2)\,(\cos 2\pi kV \cos 2\pi lW) -$$

$$(\,|\,F_{(0kl)}\,|^2 - |\,F_{(0k\bar{l})}\,|^2)\,(\sin 2\pi kV \sin 2\pi lW)\}$$

The program would then need to know the signs of the $|\,F\,|^2$ terms and the form of the trigonometric products.

Output. The output varies depending on whether a full three-dimensional Patterson synthesis is being calculated or simply a projection. In the former case, sections of vector space are taken up an axis and if, for example, it is specified that sections are to be taken at intervals of 1/30 up the axis, then thirty-one sections of output will be obtained, section 0 and section 30 being the same Each section will consist of a rectangular network of 31×31 points. At each point the numerical value of $P(XYZ)$ is printed.

If a projection is calculated for, say, the $0kl$ reflections, i.e. a projection down the a-axis, then only one layer of output is obtained. From two projections down different axes the co-ordinates of any observable peaks can be obtained in vector space.

FOURIER SYNTHESIS PROGRAM

Purpose. Unlike the Patterson synthesis, a Fourier synthesis uses the magnitudes and phases of the structure factors. A three-dimensional electron density contour map is produced, in which each atom in the unit cell should be visible (possibly excluding hydrogen atoms which contain only one electron). If the space group of the structure being considered is centrosymmetric then the phases of the structure factors are either $+$ or $-$. In the initial stages of a structure analysis employing the heavy-atom method, the co-ordinates of the heavy atom obtained from a Patterson synthesis allow phases to be allocated to every structure factor on the assumption that the scattered rays will be dominated by the effect of the heavy atom. In this way the positions of the lighter atoms in the structure can be found from a Fourier synthesis.

Data Required

(a) Once the heavy atom position has been found, a least-squares procedure can be used to obtain a more accurate position for it by calculating the best fit between the observed and calculated structure factors. The input data for the Fourier synthesis would then be the sorted magnitudes of the structure factors together with the sign of the calculated structure factors. Each structure factor has an associated set of *hkl* indices:

	hkl	$F_{(hkl)}$
	001	$+412$
e.g.	010	-60
	100	$+714$

Most Fourier programs require data to be presented in a particular order. This is because the program can be made more efficient regarding computing time if the calculations are carried out using not the general Fourier expression, but one which has been modified trigonometrically to a special form.

(b) Information is needed about the expression to be used for the Fourier summation, and this is dependent upon the space group of the structure being examined. The electron density ρ at any point X, Y, Z, in the unit cell is given by:

$$\rho(XYZ) = \frac{1}{V}\{F_{(000)} + 2\sum_{h=0}^{+\infty}\sum_{k=-\infty}^{+\infty}\sum_{l=-\infty}^{+\infty} F_{(hkl)}$$
$$\cos\left(2\pi(hX + kY + lZ) - \alpha_{(hkl)}\right)\}$$

as explained earlier.

If the structure contains a centre of symmetry then the above expression simplifies to:

$$\rho(XYZ) = \frac{1}{V}\{F_{(000)} + 2\sum_{h=0}^{+\infty}\sum_{k=-\infty}^{+\infty}\sum_{l=-\infty}^{+\infty} F_{(hkl)}$$
$$\cos 2\pi(hX + kY + lZ)\}$$

For ease of computation the cos and sin terms of the particular Fourier expression applicable to the crystal space group can be expanded using relationships such as:

(i) $\sin(A \pm B) = \sin A \cos B \pm \cos A \sin B$
(ii) $\cos(A \pm B) = \cos A \cos B \mp \sin A \sin B$

In this way, allowance may be made for the symmetry of the

space group, and the electron density formula can be expressed as a combination of product functions of the type $\cos 2\pi hX \sin 2\pi kY \cos 2\pi lZ$, whose coefficients may be combinations of $A(hkl) \pm A(\bar{h}kl) \pm A(h\bar{k}l) \pm A(hk\bar{l})$, etc.

A typical procedure that would be followed when calculating a Fourier synthesis is then, first obtain the appropriate electron density formula for the particular space group from *International Tables* Vol. I, p. 367, expand this equation so that the trigonometric terms are expressed as products, and describe the resultant expression to the computer so that the general expression may be suitably modified. For example, consider space group No. 15 $C2/c$, when hkl and $\bar{h}kl$ reflections are being used as data. The electron density formula is:

$$\rho(XYZ) = \frac{4}{V_c} \left\{ \sum\sum\sum{}^{l=2n} \left[F_{(hkl)} \cos 2\pi(hX + lZ) + F_{(\bar{h}kl)} \cos 2\pi \right. \right.$$
$$\left. (-hX + lZ) \right] \cos 2\pi kY -$$

$$\sum\sum\sum{}^{l=2n+1} \left[F_{(hkl)} \sin 2\pi(hX + lZ) + F_{(\bar{h}kl)} \sin 2\pi(-hX + lZ) \right]$$
$$\sin 2\pi kY \}$$

This expands to:

$$\rho(XYZ) = \frac{4}{V_c} \left\{ \sum{}^{l=2n} [F_{(hkl)} + F_{(\bar{h}kl)}] \cos 2\pi hX \cos 2\pi kY \right.$$
$$\cos 2\pi lZ + [-F_{(hkl)} + F_{(\bar{h}kl)}] \sin 2\pi hX \cos 2\pi kY \sin 2\pi lZ$$
$$+ \sum{}^{l=2n+1} [-F_{(hkl)} + F_{(\bar{h}kl)}] \sin 2\pi hX \sin 2\pi kY \cos 2\pi lZ$$
$$+ [-F_{(hkl)} - F_{(\bar{h}kl)}] \cos 2\pi hX \sin 2\pi kY \sin 2\pi lZ$$

The information needed by the program would be:
  (a) The signs associated with the $F_{(hkl)}$.
  (b) A description of the trigonometrical products.

Output. The form of the output is similar to that obtained for a Patterson synthesis except that in the Fourier case a three-dimensional electron density map of real space is obtained.

THE DIFFERENCE FOURIER

A difference Fourier is obtained by using as the coefficients ΔF values, where $\Delta F = |F_0| - |F_c|$.

If a structure has been satisfactorily solved and refined, then the calculation of a difference Fourier gives a measure of the correctness of the crystal structure, since the calculated structure factor moduli are based on the crystal structure model. Any atoms which have not been found should show up on the difference Fourier map.

For the purpose of the calculation the phases of the calculated structure factors for the determined structure are used, and the Fourier map that is obtained represents the difference in electron density between a Fourier synthesis using observed structure factor moduli with the phases of the calculated structure factors, and a Fourier synthesis using the calculated structure factor moduli with the calculated phases.

It is sometimes possible to determine the positions of hydrogen atoms in a structure by examining the final difference Fourier closely.

The Fourier expression that is used in calculating the difference in electron density between the postulated and observed structure may be expressed as:

$$\Delta\rho(xyz) = \frac{1}{V}\sum_h\sum_k\sum_l\left(|F_0| - |F_c|\right)e^{i\alpha}\,e^{-2\pi i(hx+ky+lz)}$$

where α is the phase associated with F_c, the calculated structure factor.

LEAST SQUARES REFINEMENT PROGRAM

Purpose. When the co-ordinates of atoms in a unit cell are known, a least squares procedure can be used in an iterative manner, to modify slightly the co-ordinates to give the best agreement between the observed structure factors and those calculated for the postulated structure. As well as refining the atomic co-ordinates, the program also refines either one isotropic (i.e. spherical) or six anisotropic (i.e. ellipsoidal) vibration parameters for each atom, and an overall scale factor.

This program may be used in the first instance to refine the heavy atom position obtained from the Patterson synthesis and later in the more important overall refinement of the complete structure. The measure of agreement between observed and calculated structure factors is related to the correctness of the final structure.

The program then must:

(a) Calculate the structure factors.

(b) Refine the various atomic parameters and the overall scale.

(c) Estimate standard deviations of the values obtained in (b).

The expression used to calculate structure factors is:

$$F(hkl)\,(\text{calc.}) = \sum_{j=1}^{N} f_j \exp\left[2\pi i(hx_j + ky_j + lz_j)\right]$$

summed over j atoms.

f_j is the atomic scattering factor corrected for atomic vibration. x, y, z, are fractional atomic co-ordinates.

It is usual to give explicitly only those fractional co-ordinates that describe non-equivalent atomic positions in the unit cell. The equivalent positions in the space group (the values of which are given in *International Tables* Vol. I) can be described in terms of R, a rotation matrix, and t, a translation vector, which the program uses to derive the co-ordinates of the equivalent atoms.

As an example of the description of the space group symmetry, consider the sodium perchlorate-bis-NN' ethylenebis(salicylidene-iminato) copper(II) complex[56]. This complex belongs to space group $C2/c$, No. 15 in *International Tables* Vol. I. There are 4 equivalent positions related by the C-face centring at $\frac{1}{2}, \frac{1}{2}, 0$, to four others. The former have co-ordinates:

$$(1)\ x, y, z;\quad (2)\ \bar{x}, \bar{y}, \bar{z};\quad (3)\ \bar{x}, y, \tfrac{1}{2}-z;\quad (4)\ x, \bar{y}, \tfrac{1}{2}+z$$

In the refinement program that was used (*NRC Crystallographic Programs* for the IBM 360 System by F. R. Ahmed, S. R. Hall & M. E. Pippy, and C. P. Saunderson) the symmetry of the general atomic positions was described in the following way:

(a) The lattice type was described by a code word, so that only the four positions described above need be considered.
(b) Positions (1) and (3) are related to (2) and (4) respectively by a centre of symmetry. The fact that the space group is centro-symmetric is also described by a code word, so that only positions (1) and (2) need to be described by the R matrix and t vector.
(c) The x, y, z, positions are described by the standard values:

$$
\begin{array}{cc}
R_1 & t_1 \\
1\,0\,0 & 0\,/\,1 \\
0\,1\,0 & 0\,/\,1 \\
0\,0\,1 & 0\,/\,1 \\
\end{array}
$$

(d) The $\bar{x}, y, \frac{1}{2}-z$ position is described by:

$$
\begin{array}{cc}
R_2 & t_2 \\
\bar{1}\,0\,0 & 0\,/\,1 \\
0\,1\,0 & 0\,/\,1 \\
0\,0\,\bar{1} & 1\,/\,2 \\
\end{array}
$$

The fractional co-ordinates of the non-equivalent atoms in the unit cell yield the related positions by the application of the relationship:

$$X^1 = Rx + t$$

In the structure being considered the sodium and chlorine atoms lie in special positions on the twofold axis, i.e. there are four of each of these atoms in the unit cell instead of the eight there would be if they were in general positions. The co-ordinates of these positions are:

$$(1)\ 0, y, \tfrac{1}{4}; \quad \text{and} \quad (2)\ 0, \bar{y}, \tfrac{3}{4}$$

as given in *International Tables* Vol. I. for this space group. For the purpose of the program they are described by allocating them a different symmetry number to the general positions. The overall representation then becomes:

Atoms	Symmetry No.	Matrices
General	1	R_1 and R_2 (plus a centre of symmetry)
Na, Cl	2	R_1 (plus a centre of symmetry)

The input data required for structure factor least squares refinement programs are then:

(a) Cell dimensions.

(b) Atomic fractional co-ordinates.

(c) An atomic vibration parameter for each atom (six vibration parameters are used for anisotropic vibrations of the atoms).

(d) A set of structure factor data for each *hkl* plane.

(e) Values of R and t to describe the space group symmetry applicable to the atoms in the unit cell.

(f) A scale factor to put the structure factors on an absolute scale. This is obtained either from a Wilson plot (see earlier) or by using an estimated value in the first refinement cycle, and allowing the program to refine it to a more realistic value for the second cycle.

(g) A set of atomic scattering factors given as values for various increments of $\sin \theta / \lambda$ for each atom, suitably corrected for the real part of the anomalous dispersion. Values of scattering factors can be found in *International Tables* Vol. III (1962), p. 207.

(h) A weighting scheme is usually included and in the later stages of refinement may be modified to ensure that the average value of $W\Delta^2$ ($\Delta = |F_0| - |F_c|$) is approximately constant for groups of structure factors arranged in order of magnitude of $|F_0|$.

Output. Outputs vary depending on the programs used, but in general the following information is obtained:

(a) A list of F_0, F_c and ΔF for each plane. (Optional)

(b) A list of shifts in atomic co-ordinates and vibration parameters to more reasonable values.

(c) The new atomic co-ordinates and vibration parameters, together with their standard deviations.

(d) A new scale factor.

(e) A calculated agreement index, R (not to be confused with the matrix, R). This should become progressively smaller as the refinement continues:

$$R = \Sigma W(|F_0| - |F_c|)^2 / \Sigma |F_0|$$

W = a weight = $1/\sigma^2$, where σ^2 = the variance.

(f) A predicted value of R for the next cycle.

(g) A list of $W\Delta^2$ for increasing values of F_0 and $\sin \theta/\lambda$.

It is usual for refinement programs to allow a choice of weighting scheme, so that suitable weights may be allocated in most cases.

INTERATOMIC DISTANCES AND ANGLES PROGRAM

Purpose. A measure of the correctness of a structure determination is provided by the agreement found between the calculated bond lengths and angles with those values accepted as normal by previous experience. For example, the C–C distance in an aromatic ring given in *Interatomic Distances*, Chemical Society Special Publication No. 11 (1958) has an average value of $1 \cdot 395 \pm 0 \cdot 003$ Å. Consequently in any other structure containing an aromatic ring one would expect an interatomic C–C distance near this value.

It is necessary, therefore, to have a program which will calculate the distance from an atom to its nearest neighbours, both where a bond is present and where Van der Waal's contacts exist. The angle between any three atoms is also calculated, and the standard deviations in the distances and angles are calculated from the standard deviations in the atomic co-ordinates obtained from the final cycle of refinement.

The distance between two atoms A and B having co-ordinates x, y, z; and x_1, y_1, z_1; is given by:

$$d = [(x-x_1)^2 + (y-y_1)^2 + (z-z_1)^2]^{\frac{1}{2}}$$

when the reference axes are an orthogonal set.

In the case of monoclinic and triclinic crystal systems the co-ordinates of atoms in the oblique cells are usually first converted to co-ordinates referred to orthogonal axes. e.g. F. R. Ahmed (*NRC Crystallographic Programs* for the IBM 360 System) uses the relationship:

$$X' = T.X \text{ Å}$$

where X' is along the a-axis, Y' is in the ab plane, and Z' is along the c^* axis.

Then:
$$T = \begin{bmatrix} ab.\cos\gamma & c.\cos\beta \\ 0b.\sin\gamma & c(\cos\alpha - \cos\beta.\cos\gamma)/\sin\gamma \\ 00 & V/(ab.\sin\gamma) \end{bmatrix}$$

Standard Deviations. It has been shown [117] that if l is the bond length between two independent atoms having variances $\sigma^2(A)$ and $\sigma^2(B)$ in the direction of the bond, then:

$$\sigma^2(l) = \sigma^2(A) + \sigma^2(B)$$

If β is the angle formed at B between two bonds AB and BC, then:

$$\sigma^2(\beta) = \frac{\sigma^2(A)}{AB^2} + \sigma^2(B)\left(\frac{1}{AB^2} - \frac{2\cos\beta}{AB.BC} + \frac{1}{BC^2}\right) + \frac{\sigma^2(C)}{BC^2}$$

F. R. Ahmed uses the following expression to obtain the standard deviation in a bond length from the atomic co-ordinates of the atoms forming the bond:

$$\sigma^2(AB) = \left\{ \begin{aligned} &(\Delta X')^2[\sigma^2(X_{A'}) + \sigma^2(X_{B'})] + \\ &(\Delta Y')^2[\sigma^2(Y_{A'}) + \sigma^2(Y_{B'})] + \\ &(\Delta Z')^2[\sigma^2(Z_{A'}) + \sigma^2(Z_{B'})] \end{aligned} \right\} \Big/ (AB)^2$$

$\Delta X'$, $\Delta Y'$, and $\Delta Z'$ are the differences
$(X_{A'} - X_{B'})$, $(Y_{A'} - Y_{B'})$, $(Z_{A'} - Z_{B'})$ respectively.

Data Required
  (a) Unit cell dimensions in Å.
  (b) Fractional co-ordinates of the asymmetric unit of the unit cell: ideally a complete molecule.
  (c) Standard deviations of the atomic co-ordinates.
  (d) A limit is usually put on the length of interatomic distance to be calculated, e.g. 3·5 Å.
  (e) Symmetry information to describe the asymmetric unit and its relationship to the rest of the cell contents.
  (f) In addition to (e), further information to describe the number of unit cells to be considered in the bond search routine.

Output. A systematically labelled output is obtained giving:

(a) Bond lengths and angles in the asymmetric unit.
(b) Interatomic distances and angles between symmetry related units.
(c) Standard deviations in the distances and angles.

Bibliography

ARNDT, V. W. and WILLIS, B. T. M., *Single Crystal Diffractometry*, Cambridge University Press (1966).

BRAGG, SIR L., *The Crystalline State*, Vol. I 'A General Survey', Bell

BUERGER, M. J., *X-ray Crystallography*, Wiley, New York (1942).

BUERGER, M. J., *Crystal Structure Analysis*, Wiley, New York (1960).

BUERGER, M. J., *Vector Space*, Wiley, New York (1959).

BUNN, C. W., *Chemical Crystallography*, Clarendon Press, Oxford (1961).

HARTSHORNE, N. H., and STUART, A., *Crystals and the Polarising Microscope*, Arnold (1960)

HENRY, N. F. M., LIPSON, H., and WOOSTER, W. A., *The Interpretation of X-ray Diffraction Photographs*, Macmillan, London (1951).

JAMES, R. W., *The Crystalline State*, Vol. II 'The Optical Principles of the Diffraction of X-rays', Bell.

LIPSON, H., and COCHRAN, W., *The Crystalline State*, Vol. III 'The Determination of Crystal Structures', Bell.

PHILLIPS, F. C., *An Introduction to Crystallography*, Longmans, London (1963).

RAMACHANDRAN, G. N. (Ed.), *Advanced Methods of Crystallography*, Academic Press, New York (1964).

ROLLETT, J. S. (Ed.), *Computing Methods in Crystallography*, Pergamon Press, Oxford (1965).

STOUT, G. H. and JENSON, L. H., *X-ray Structure Determination*, Macmillan, New York (1968).

WOOLFSON, M. M., *Direct Methods in Crystallography*, Clarendon Press, Oxford.

International Tables for X-ray Crystallography, Vol. I, II and III, Kynoch Press (1952–1962).

References

1. BUERGER, M. J., *X-ray Crystallography*, Wiley, New York, 4 (1942)
2. PHILLIPS, F. C., *An Introduction to Crystallography*, 2nd edn, Longmans, London, 40 (1963)
3. PHILLIPS, F. C., *An Introduction to Crystallography*, 2nd edn, Longmans, London, 226 (1963)
4. *International Tables for X-ray Crystallography*, I, 23
5. *International Tables for X-ray Crystallography*, I, 25
6. *International Tables for X-ray Crystallography*, I, 50
7. BUNN, C. W., *Chemical Crystallography*, Oxford University Press, 111 (1961)
8. BUERGER, M. J., *X-ray Crystallography*, Wiley, New York, 178 (1942)
9. ALCOCK, N. W., SHELDRICK, G. M., *Acta Cryst.*, **23**(1), 35 (1967)
10. PHILLIPS, F. C., *An Introduction to Crystallography*, 2nd edn, Longmans, London, 282 (1963)
11. *International Tables for X-ray Crystallography*, **III**, 157; BUERGER, M. J., *X-ray Crystallography*, Wiley, 181
12. MILBURN, G. H. W., TRUTER, M. R., *J. Chem. Soc.* (A), 1609 (1966)
13. HARTSHORNE, N. H. and STUART, A., *Crystals and the Polarising Microscope*, Arnold, London, 270 (1960)
14. HENDERSHOT, O. P., *Rev. Sci. Instrum.*, **8**, 436 (1937)
15. WEISZ, O., COLE, W. F., *J. Sci. Instrum.*, **25**, 213–214 (1948)
16. BUNN, C. W., *Chemical Crystallography*, Oxford University Press, 155 (1961)
17. KRATKY, O., KREBS, B., *Z. Kristallogr.*, **95**, 253 (1937)
18. BUNN, C. W., *Chemical Crystallography*, Oxford University Press, 149 (1961)
19. BUNN, C. W., *Chemical Crystallography*, Oxford University Press, 155 (1961)
20. BUNN, C. W., *Chemical Crystallography*, Oxford University Press, 180 (1961)
21. *Index (Inorganic) to the Powder Diffraction File*, Am. Soc. for Testing and Materials (1963)
22. BUNN, C. W., *Chemical Crystallography*, Oxford University Press, 198 (1961)
23. BUERGER, M. J., *X-ray Crystallography*, Wiley, New York, 255 (1942)
24. The Institute of Physics and the Physical Society, 47 Belgrave Square, London, S.W.1.
25. BUNN, C. W., *Chemical Crystallography*, Oxford University Press, 257 (1961)
26. *International Tables for X-ray Crystallography*, **I**, 348
27. BUERGER, M. J., *Crystal Structure Analysis*, Wiley, New York, 79 (1960)
28. ROBERTSON, J. H., *J. Sci. Instrum.*, **37**, 41 (1960)
29. STOUT, G. H. and JENSEN, H., *X-ray Structure Determination*, Macmillan, New York 122 (1968)
30. BUERGER, M. J., *The Precession Method*, Wiley, New York (1964)
31. COOPER, A., FRASER, K. A., FREEMAN, H. C., *Operation Manual for the Sydney University Supper Diffractometer*. See also FREEMAN, H. C., GUSS, J. M., NOCKOLDS, C. E., PAGE, R. and WEBSTER, A., *Acta Cryst.*, A26, 149 (1970)
32. WOOSTER, W. A., *J. Sci. Instrum.*, **42**, 9, 685 (1965)
33. STOUT, G. H. and JENSEN, H., *X-ray Structure Determination*, Macmillan, 185 (1968)
34. ARNDT, V. W. and WILLIS, B. T. M., *Single Crystal Diffractometry*, Cambridge University Press, 220 (1966)

35. ARNDT, V. W. and WILLIS, B. T. M., *Single Crystal Diffractometry*, Cambridge University Press, 270 (1966)
36. STOUT, G. H. and JENSEN, H., *X-ray Structure Determination*, Macmillan, New York, 454 (1968)
37. *The World List of Crystallographic Computer Programs*, 2nd edn, International Union of Crystallography
38. JOHNSON, C. K., 'ORTEP: A FORTRAN Thermal Ellipsoid Plot Program for Crystal Structure Illustrations'. *Oak Ridge National Laboratory Report*, ORNL—3794 (1965)
39. PHILLIPS, D. C., *Acta Cryst.*, **7,** 746 (1954)
    PHILLIPS, D. C., *Acta Cryst.*, **9,** 819 (1956)
40. BUSING, W. R. and LEVY, H. A., *Acta Cryst.*, **10,** 180 (1957)
41. WELLS, M., *Acta Cryst.*, **13,** 722 (1960)
42. EVANS, H. T. and ECKSTEIN, M. G., *Acta Cryst.*, **5,** 540 (1952)
43. BRADLEY, A. J., *Proc. Phys. Soc.*, **47,** 879 (1935)
44. STOUT G. H. and JENSEN, H., *X-ray Structure Determination*, Macmillan, New York, 409 (1968)
45. DARWIN, C. G., *Phil. Mag.*, **43,** 800 (1922)
46. LONSDALE, K., *Min. Mag.*, **28,** 14–25 (1947)
47. ZACHARIASEN, W. H., *Acta Cryst.*, **16,** 1139 (1963)
48. ASBRINK, S. and WERNER, P. E., *Acta Cryst.*, **20**(3), 407 (1966)
49. PATTERSON, A. L., *Phys. Rev.*, **46,** 372 (1934)
50. PATTERSON, A. L., *Z. Kristallogr.*, **90,** 517 (1935)
51. BUERGER, M. J., *Vector Space*, Wiley, New York, 19 and 25 (1959)
52. BUERGER, M. J., *Vector Space*, Wiley, New York, 199 (1959)
53. HARKER, D., *J. Chem. Phys.*, **4,** 381 (1936)
54. BUNN, C. W., *Chemical Crystallography*, Oxford University Press, 416 (1961)
55. FREEMAN, H. C. and MILBURN, G. H. W., unpublished work
56. MILBURN, G. H. W., TRUTER, M. R., and VICKERY, B. L., *Chem. Comm.*, 1188 (1968)
57. JACOBSON, R. A., *The Symmetry Map and Vector Verification*, Lecture Notes, NATO Advanced Study Institute, Direct and Patterson Methods, University of York, England, September (1971)
58. MIGHELL, A. D. and JACOBSON, R. A., *Acta Cryst.*, **16,** 1046 (1963)
59. BRAUN, P. B., HORNSTA, J. and LEENHOUTS, J. I., *Phil. Res. Tech. Notes*, 123, 124 (1968)
60. HUBBARD, C. R. and JACOBSON, R. A., *Ames. Lab. Report* IS-2210 (1969)
61. HUBBARD, C. R., QUICKSALL, C. O. and JACOBSON, R. A., *Ames Lab. Report* IS-2625 (1971)
62. MIGHELL, A. D. and JACOBSON, R. A., *J. Natl. Bur. Stand.*, 70A, 319 (1966); RAMAN, S. and KATZ, J. L., *Z. Krist.*, **124,** 43 (1967)
63. HACKERT, M. L. and JACOBSON, R. A., *Acta Cryst.*, B26 1682 (1970)
64. TOLLIN, P., *Patterson Function Interpretation Techniques*, Lecture Notes, NATO Advanced Study Institute, Direct and Patterson Methods, University of York, England, September 1971.
65. ROSSMANN, M. G. and BLOW, D. M., *Acta Cryst.*, **15,** 24 (1962)
66. TOLLIN, P., MAIN, P., ROSSMANN, M. G., *Acta Cryst.*, **20,** 404 (1966)
67. BRAUN, P. B., HORNSTRA, J., LEENHOUTS, J. I., *Phil. Res. Repts.*, **24,** 85 (1969)
68. NORDMAN, C. E., *Trans. Amer. Cryst. Assn.*, **2,** 29 (1966); NORDMAN, C. E., *Cryst. Computing*, Munksgaard (1970)
69. HOPPE, W., *Z. Electrochem.*, **61,** 1076 (1957)
    HOPPE, W., *Z. Krist.*, **117,** 4 (1970)
70. BUERGER, M. J., *Advanced Methods of Crystallography*, Ed. G. N. Ramachandran, Academic Press, New York (1964)

71. TOLLIN, P. and COCHRAN, W., *Acta Cryst.*, **17**, 1332 (1964); TOLLIN, P., *Acta Cryst.*, **21**, 613 (1966)
72. CROWTHER, R. A. and BLOW, D., *Acta Cryst.*, **23**, 544 (1967)
73. HOPPE, W., *et al.*, *Angew. Chemie.*, **6**, 809 (1967)
74. SAYRE, D., *Acta Cryst.*, **6**, 430 (1953)
75. LIPSON, H. and COCHRAN, W., *The Determination of Crystal Structures*, Bell, 11
76. LIPSON, H. and COCHRAN, W., *The Determination of Crystal Structures*, Bell, 77
77. CRUICKSHANK, D. W. J., *The Equations of Structure Refinement*, private communication
78. BIJFOET, J. M., *Proc. Koninkl. Ned. Akad. Wetenschap.*, (B)52, 313 (1949)
79. PEERDEMAN, A. F., V. BOMMEL, A. J. and BIJFOET, J. M., *Proc. Koninkl. Med. Akad. Wetenschap*, (B)54, 16 (1951)
80. BIJFOET, J. M., *Nature*, **173**, 888 (1954)
81. TRAMMEL, J. and BIJFOET, J. M., *Acta Cryst*, **7**, 703 (1954)
82. BIJFOET, J. M., *Endeavor*, **14**, 71 (1955)
83. RAMASESHAN, S., *Advanced Methods of Crystallography*, Ed. G. N. Ramachandran, Academic Press, New York, 67 (1964)
84. OKAYA, Y. and PEPINSKY, R., *Computing Methods and the Phase Problem in X-ray Crystal Analysis*, Ed. R. Pepinsky, J. M. Robertson, J. C. Speakman, Pergamon Press, Oxford, 273 (1961)
85. NOCKOLDS, C. K., RAMASESHAN, S., WATERS, T. N. M., WATERS, J. M., CROWFOOT HODGKIN, D., *Nature*, **214**, 129 (1967)
86. LIPSON, H. and COCHRAN, W., *The Determination of Crystal Structures*, Bell, 61
87. WILSON, A. J. C., *Nature*, **150**, 151 (1942)
88. STOUT, G. H. and JENSEN, H., *X-ray Structure Determination*, Macmillan, New York, 449 (1968)
89. LEGENDRE, A. M., *Nouvelles methodes pour la determination des orbites des cometes*, Courcier, Paris, 72 (1806)
90. CRUICKSHANK, D. W. J., *Acta Cryst.*, **13**(2), 775 (1960)
91. STOUT, G. H. and JENSEN, H., *X-ray Structure Determination*, Macmillan, New York, 390 (1968)
92. ARNDT, U. W. and WILLIS, B. T. M., *Single Crystal Diffractometry*, Cambridge University Press, 269 (1966)
93. MILBURN, G. H. W. and TRUTER, M. R., *J. Chem. Soc. (A)* 648 (1967)
94. BUERGER, M. J., *Crystal Structure Analysis*, Wiley, New York, 262 (1960)
95. HARKER, D. and KASPER, J. S., *Acta Cryst.*, **1**, 70 (1948)
96. WOOLFSON, M. M., *Direct Methods in Crystallography*, Oxford University Press
97. BUERGER, M. J., *Crystal Structure Analysis*, Wiley, New York, 557 (1960)
98. WOOLFSON, M. M., *Direct Methods in Crystallography*, Oxford University Press, 25
99. SAYRE, D., *Acta Cryst.*, **5**, 60 (1952)
100. WOOLFSON, M. M., *Direct Methods in Crystallography*, Oxford University Press, 49
101. COCHRAN, W. and WOOLFSON, M. M., *Acta Cryst.*, **8**, 1 (1955)
102. WOOLFSON, M. M., *Direct Methods in Crystallography*, Oxford University Press 52
103. ZACHARIASEN, W. H., *Acta Cryst.*, **5**, 68 (1952)
104. WOOLFSON, M. M., *Direct Methods in Crystallography*, Oxford University Press, 55
105. COCHRAN, W., *Acta Cryst.*, **5**, 65 (1952)
106. KARLE, J. and KARLE, I. L., *Acta Cryst.*, **21**, 849 (1966)
107. HAUPTMAN, H. and KARLE, J., *A.C.A. Monograph*, **3**, (1953)
108. KARLE, J. and KARLE, I. L., *Acta Cryst.*, **21**, 860 (1966)

109. WOOLFSON, M. M., *Direct Methods in Crystallography*, Oxford University Press, 88
110. GILLIS, J., *Acta Cryst.*, **1,** 174 (1948)
111. WOOLFSON, M. M., *Acta Cryst.*, **7,** 61 (1954)
112. KARLE, J. and KARLE, I. L., *Acta Cryst.*, (B)**24,** 81 (1968)
113. AHMED, F. R. and HALL, S. R., Private communication. *NRC Crystallographic Programs* for the IBM 360 System
114. GERMAIN, G. and WOOLFSON, M. M., *Acta Cryst.*, (B)**24,** 91 (1968)
115. RABINOWITZ, I. N. and KRAUT, J., *Acta Cryst.*, **17,** 159 (1964)
116. LAING, M., *Acta Cryst.*, B25(8), 1674 (1969)
117. CRUICKSHANK, D. W. J. and ROBERTSON, J., *Acta Cryst.*, **6,** 698 (1953)

Index

213